Electrical Installation Practice

Electrical Installation Practice Book 2

Fourth edition

H. A. Miller

Revised by
R. D. Puckering LCG, BEd Hons
Paddington College

BSP PROFESSIONAL BOOKS

OXFORD LONDON EDINBURGH

BOSTON MELBOURNE

Copyright © R. D. Puckering 1990

All rights reserved. No part of this
publication may be reproduced, stored
in a retrieval system, or transmitted,
in any form or by any means, electronic,
mechanical, photocopying, recording
or otherwise without the prior
permission of the copyright owner.

First published 1990

British Library
Cataloguing in Publication Data
Miller, Henry A
  Electrical installation practice—4th ed.
  Bk. 2
  1. Electrical equipment. Installation
  I. Title II. Puckering, R. D.
  621.31'042

ISBN 0-632-02541-7

BSP Professional Books
A division of Blackwell Scientific
  Publications Ltd
Editorial Offices:
Osney Mead, Oxford OX2 0EL
  (Orders: Tel. 0865 240201)
25 John Street, London WC1N 2BL
23 Ainslie Place, Edinburgh EH3 6AJ
3 Cambridge Center, Suite 208, Cambridge,
  MA 02142, USA
107 Barry Street, Carlton, Victoria 3053,
  Australia

Set by Setrite Typesetters Ltd

Printed in Great Britain at the
  University Press, Cambridge

Contents

Preface vii

Acknowledgements viii

One The Process of Installation 1
1.1 The roles of members of the design team 1
1.2 The roles of members of the construction team 3
1.3 Professional and contractual relationships 4
1.4 The company structure 4
1.5 Installation team 6
1.6 Planning the installation 8
Test 1 16

Two Electric Cables and Factors Determining Their Choice 17
2.1 Electric cables 17
2.2 Factors determining choice of cable 19
2.3 The selection of conductor size 22
Test 2 26

Three Protection Against Indirect Contact 27
3.1 Earthing 27
3.2 Equipotential bonding 29
3.3 Supplementary bonding 31
3.4 Circuit protective conductors 32
3.5 Connecting protective conductors and earth electrodes 33
3.6 Protection by automatic disconnection from the supply 36
3.7 Residual current devices 37
Test 3 39

Four Control and Protection for the Consumer 40
4.1 Consumer's switchgear 40
4.2 IEE Regulations concerning switchgear 43
4.3 Protection of the installation 44
Test 4 51

Five The Installation of Cable Tray and Ladder Rack 52
5.1 Cable tray 52
5.2 Determining the size of cable tray 52
5.3 Types of cable tray 52
5.4 Installation of cable tray 56

5.5	Ladder racking	59
	Test 5	60

Six The Installation of Cable Trunking and Under Floor Ducts — 61
- 6.1 Metallic trunking — 61
- 6.2 Determining the size — 61
- 6.3 Installation of metallic cable trunking — 62
- 6.4 Types of metallic cable trunking — 65
- 6.5 Under floor trunking and cable ducts — 68
- 6.6 Bus-bar trunking systems — 71
- 6.7 Plastic trunking — 73
- 6.8 Multi-compartment trunking — 74
- 6.9 Segregation of circuits — 74
- 6.10 Electromagnetic effects — 75
- Test 6 — 76

Seven The Heating of Water by Electricity — 77
- 7.1 Water heating — 77
- 7.2 IEE Regulations concerning water heating — 83
- 7.3 Electric heating calculations — 84
- Test 7 — 87

Eight Cooking and Space Heating — 88
- 8.1 The installation of cooker final circuits — 88
- 8.2 Control and protection of cooker circuits — 88
- 8.3 The application of diversity to cooker circuits — 90
- 8.4 Heating — 92
- 8.5 Electric heating — 94
- 8.6 Space heating — 100
- 8.7 IEE Regulations concerning space heating — 102
- Test 8 — 104

Nine Electric Lighting — 105
- 9.1 The incandescent lamp — 105
- 9.2 The tungsten filament lamp — 106
- 9.3 The tungsten halogen lamp — 108
- 9.4 Fluorescent lighting — 109
- 9.5 Stroboscopic effects — 113
- 9.6 High voltage discharge lighting — 114
- 9.7 Emergency lighting — 119
- Test 9 — 123

Ten The Inspection and Testing of Installations — 124
- 10.1 Inspecting and testing — 124
- 10.2 The testing of installations — 124
- 10.3 Certification — 130
- Test 10 — 132

Answers to the Tests — 133

Index — 135

Preface

This well known series of books on the craft practice aspects of electrical installation work has been updated to meet the new emphasis on practical competence. The books cover installation practice, associated craft theory, safe working practice, and a study of the electrical industries.

Book 2 takes you through the second half of the City and Guilds 236 Part 1 course. It goes more deeply into some of the work covered by Book 1 and then introduces you step by step to new topics covered by the syllabus.

It is recognised that students for the Part 1 certification in Electrical Installation will have come from very varied backgrounds, and will have different employment and training experiences. So this book, in accordance with the requirements of the syllabus, starts out by attempting to explain in simple terms the roles that the various people involved in the industry play, and the relationships both professional and contractual between them.

Following chapters explain in a practical way the installation of cable tray, ladder rack and cable trunking, and give reasons for the choice of these and other wiring systems, together with the rules and regulations concerning their use. The installation of water heating, cooker circuits, space heating and discharge lighting equipment is discussed, together with the calculations required to ensure their safe and effective application.

The book reflects the emphasis placed on safety in the latest edition of the IEE Wiring Regulations and the installation of circuit protective conductors and circuit protective devices are fully discussed. All explanations have clear illustrations to guide you through the correct procedures, and you can check your knowledge with the multi-choice questions at the end of each chapter.

R. D. Puckering

Acknowledgements

British Insulated Callander Cables Ltd
Central Generating Board
Chloride Gent
J. A. Crabtree & Co Ltd
Davis Trunking Ltd
Delta Accessories and Domestic Switchgear Ltd
The Institute of Electrical Engineers
Martindale
Midland Electrical Manufacturing Company Ltd
M. K. Electric Ltd
Pyrotenax Ltd
Rawlplug Ltd
Swifts of Scarborough Ltd
Wylex (Scholes)

Chapter 1
The Process of Installation

1.1 The roles of members of the design team

The client

The most important person in the construction or refurbishment of a building is the client. This is the person or organisation who has placed an order for the work to be carried out, and who at the end of the contract will have paid for it to be completed. The client's requirements will dictate to a large extent the form that the building will take; however, unless the client is a large organisation employing its own design team, the details of the construction will be entrusted to a professional architect.

The Architect

Architects are skilled in the design and drawing up of plans for the fabric of the building. Their role is to advise the client on the practicality of their wishes, and to try to come up with a scheme that not only satisfies the client from a practical and aesthetic point of view, but complies with the many rules and regulations involved in the construction of buildings. While for a very small project the Architect may feel qualified to draw up plans for a complete scheme, on larger buildings he will consult specialist design engineers about technical details such as building services.

Consultants

Because the design for the larger project will be quite intricate it will require the skills of someone who has studied the problems of specific services such as electrical installation, heating and ventilation etc. to design these aspects of the project. These designers are professional people and are known as Consultant Engineers. Their role is to study the requirements of the client and the Architect's interpretation of these, and to come up with a design that will satisfy not only the client's requirements, but also on an electrical installation the requirements of the Institute of Electrical Engineers (IEE) Wiring Regulations, British Standards Specifications and the requirements of the Supply Company. They will also among other things be responsible for ensuring that cable sizes have been calculated correctly, that the capacities of the cable trunking and conduit are adequate and that the protective devices are rated correctly.

The design team must produce drawings, schedules and specifications to be sent out to the companies tendering for the contract and answer any queries that might arise from this. When the contract has been placed, they will produce drawings showing any amendments to the original tender drawings, and act as a link between the client, the main contractor and the installation team.

Quantity Surveyor

The Quantity Surveyor, or QS, has a varied role. In the context of the construction site the QS is responsible to the Architect for the measurement of all the materials that go to make up the 'bill of quantities' for the job. This contains details of building materials required, so that builders tendering for the contract have details from which they can prepare their estimates. Another and more important role as far as the electrical sub-contractor is concerned, is that the QS is responsible for measuring the materials of any additional work that may be carried out during the course of the contract and also for checking that interim claims for work done are accurate.

Clerk of Works

The role of the Clerk of Works is to ensure on behalf of the Architect that the installed materials meet the standards laid down in the specification and drawings for the job both for quality and workmanship. On larger contracts there may even be specialist Clerks of Works for electrical and/or heating and ventilation. The Clerk of Works will want to inspect parts of the job during the contract and witness any tests carried out. He may be given the authority by the Architect to sign daywork sheets and to issue the Architect's Instructions (A.I.) for alterations or additional work.

Design and construct

The above make up the members of the Architect's design team. However, this is not the only way that installations are designed. On a very small job the client will approach the electrical contractor direct and ask him/her to carry out the work. The client in this case will provide a few details but will otherwise expect the electrician to see that the installation is adequately designed.

Larger companies and in particular those that are multi-disciplined, that is to say they have engineers dealing with all the building services, often offer a design service for customers. This has the advantage that the client simply gives them the details and they do the rest. The disadvantage is that the services of the Architect are lost and in any dispute over design aspects of the job there is no one to arbitrate.

1.2 The roles of members of the construction team

The main contractor

A large number of the contracts carried out by companies in the electrical installation industry are done on construction sites. For the most part the electrical contractor will be one of a number of different sub-contractors employed by the main contractor. The main contractor, or general contractor, is usually the builder, although on sites such as refurbishment work, where the builder's work is small, the electrical contractor may well be the main contractor. It is the main contractor's job to act as co-ordinator of all the individual sub-contractors employed on the site and to see that they carry out their job to the specification and drawings and in the time allowed.

Sub-contractors

There are two types of sub-contractor, as described here.

Nominated sub-contractors These are contractors who in the opinion of the Consultant and/or the Architect are suitable contractors to carry out this particular work. That is to say they are experienced in this type of work or are specialists in it. This is important for some types of contract, for example a company whose main business was rewiring houses would in most cases be unsuitable for carrying out the installation of say a hospital. Although nominated by the Consultant or Architect the nominated sub-contractors will still in most cases have to prepare a competitive tender and sign a contract with the main contractor to complete their part of the work.

Sub-contractors These are usually appointed by the builder, sometimes by competitive tender or sometimes just invited to carry out the work. The Architect will have little say in the selection of the sub-contractor, relying on the main contractor's judgement in the matter.

If part of the work on a large contract calls for a certain specialist knowledge, then with the permission of the Architect sub-contractors can employ specialist companies to do this part of the work. Examples of this might be the installation of lifts or escalators or a closed circuit television system. In cases such as this the specialist sub-contractor signs a contract with the electrical sub-contractor and becomes what is known as a sub-sub-contractor.

Suppliers

Below are the two main types of supplier of goods and equipment to a site.

Nominated suppliers These are chosen by the Architect and/or Consultant. They will have been chosen to supply particular pieces of equipment which are required for the contract. Examples of this would be lighting fitting manufacturers who will supply a specific type of luminaire, or say the manufacturer of door furniture which the client desires to be used in the building. The equipment will have to be supplied by these people and no other type of luminaire or door furniture can be utilised.

Supplier This will be a Factor, Wholesaler or Builder's Merchant. The materials they supply will be of the sort that is readily available and, provided it meets the specifications laid down for it, can be obtained from any supplier.

1.3 Professional and contractual relationships

The relationship between the above members of the contracting industry appears intricate to the novice; Information Sheet No. 1A shows how these interrelate. The relationships fall into two main parts, as follows.

Professional relationships

These exist between the Architect and the client; the Architect will have been asked to undergo the organisation and design of the project in a professional capacity. Likewise the Architect will have asked the Consultants to participate in the design of the project, but their relationship will not be a contractual one in the strictest sense but a professional one. However, it is worth noting at this point that demands for 'quality assurance' with regards to aspects of design for construction and building services are now coming into force and things are changing in this respect for all professional people. Both the Architect and the Consultants will have an advisory role to play when the project gets under way and they will have contact with the main contractor and the sub-contractors in a professional capacity.

Contractual relationship

These exist between the client and the main contractor. The main contractor will have signed a contract with the client for erection and completion of the project. This is essential so that, if anything goes wrong with the work, each of the two parties knows where they stand in the eyes of the law. Sub-contractors in their turn sign contracts with the main contractor to carry out their part of the project, as do nominated suppliers of equipment. The diagram in Information Sheet No. 1A shows both the professional and contractual relationships between the various members of the construction industry.

1.4 The company structure

Company structures vary between different firms, but basically they fall into two main types, described below.

The Process of Installation 5

Information Sheet No. 1A Contractual and professional relationships.

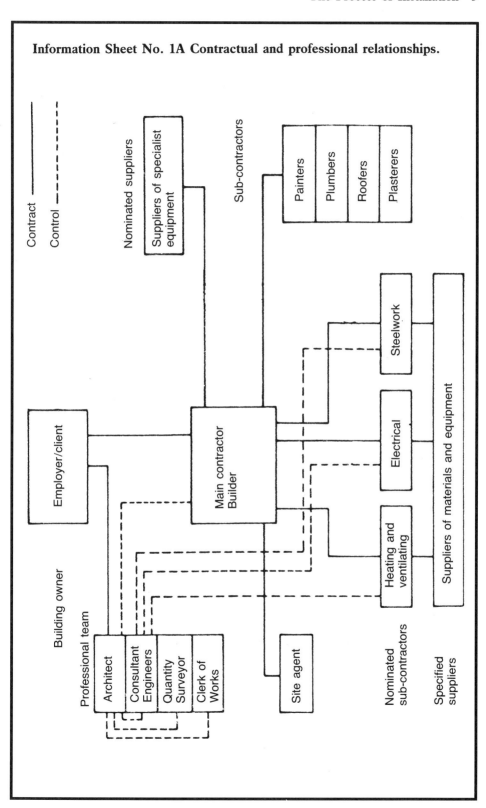

Vertical structure

In the vertical structure the various tasks, such as dealing with the initial enquiry, estimating, supervision of the contract, financial control, and the settling of the final account for a particular contract, are the responsibility of one person.

The advantage of this system is that most of the work is being dealt with by one person so that the lines of communication within the organisation are very short. People from outside bodies know who to get in touch with about a particular project, and they are not passed from one department to another.

The disadvantage of this type of system is that if the person dealing with the project is ill or leaves the company for some reason, all continuity is lost and it is extremely difficult for someone else to pick up where the other person left off.

Horizontal structure

In this system the various tasks are divided between different people such as estimators, who will deal with the enquiry and produce an estimate for a particular project, and contract supervisors, who if the tender for a particular contract is successful will have the task of implementing it. This has the advantage that the individuals become very proficient in their particular specialism and therefore work more efficiently.

The disadvantage in this system of working is that the lines of communication between the different sections become extended and a certain degree of flexibility is lost. Clients are not always sure who to speak to regarding certain aspects of a particular project and can be passed from one person to another before finally getting an answer to their problem.

1.5 Installation team

If a company is successful and wins a contract it will be notified by the Architect, Consultant or specifying authority. If the company is small it may be given to the engineer who prepared the original estimate; with a larger company it will be given to the contract engineer who is going to supervise the project. The work will have to be 'pre-planned' in accordance with certain requirements:

- The builders programme;
- Length of contract;
- Availability of labour;
- Delivery dates for certain materials;
- Access to site;
- Hire of plant and equipment;
- Health and Safety requirements.

If the contract is on an existing building then a site visit will be useful so that cable runs can be planned and measurements taken. However if the building has not been erected then this will have to be done taking measurements from the drawings. Materials will have to be ordered in good time, especially those on long delivery dates; if the company employs a buyer he will direct the Contracts Engineer as to which company offers the most advantageous discounts. The actual delivery of materials must be planned so that they spend the minimum amount of time on site to lessen the risk of damage. Fragile and/or expensive equipment must be placed in a secure store to avoid damage and theft. The Contracts Engineer will be anxious to find out which foreman has been appointed to this particular contract, as early contact can be beneficial to both of them. Contracts Engineers often run several projects at any one time and so they rely on a good relationship with site personnel in order to keep things running smoothly. It will be seen from the chart in Information Sheet No. 1A that although there is no contractual relationship between the electrical sub-contractor and the Consultant/Architect there is a professional one. Some contracts are highly technical and there will be a good deal of liaison required between the Contracts Engineer and the designer; however, it should be stressed here that from a contractual point of view the sub-contractor works for the main contractor and any decisions made should go through them.

On site the project will be in the hands of a site supervisor or foreman. Experienced in electrical installation work, this person will be responsible for the supervision of the approved electricians, apprentices and labourers carrying out the work, not only for the quality of the work they produce but for their safety and welfare as well. Records will be kept of the hours spent by each individual on site, and time sheets signed at the end of each week. The site amenities will be checked to see if these meet the requirements of the Health and Safety at Work Act and if accidents do occur these will be logged in the accident book.

The foreman or supervisor will be the company's representative on site and will attend site meetings with the Contracts Engineer and liaise with the main contractor, Clerk of Works, other sub-contractors and, through the main contractor, members of the design team. Records will be kept of any requests for alterations to the contract and details of these sent to head office. These alterations will be noted on a set of drawings known as the 'as fitted drawings' which will be required to be handed over to the client on completion of the contract.

Another important task of the foreman or supervisor is to ensure that goods and equipment, including those on hire, arrive at the times that they are required. This often means getting in touch with suppliers and chasing outstanding materials. When equipment on hire is finished with, no time should be wasted in returning it as the additional costs incurred in keeping it longer than necessary can be enormous.

Finally it is recommended that a site diary is kept in which all the above incidents are noted. This can be invaluable if later on in the contract disputes of one nature or another arise and the foreman or supervisor is able to show details of what really transpired.

The City and Guilds of London Institute have in recent years adopted the use of assignments as part of the student assessment. These assignments incorporate a good deal of the above planning and preparation of material lists and are invaluable as a means of experiencing some of the problems encountered when running electrical contracts. We shall therefore look at this aspect of the course in more detail.

1.6 Planning the installation

Drawings and specification

The first concept of what the electrical installation is to be like is gained by the electrician when looking at the specification and drawings for the first time. In simple terms the specification will give him details of the type and quality of the materials and the standards of workmanship required, while the drawing will show him details of the location of all the items of equipment and enable him to determine the quantity of materials required.

The types of drawings most commonly encountered by the electrician are:

- Circuit diagrams;
- Wiring diagrams;
- Block diagrams;
- Layout drawings;

Circuit diagrams A circuit diagram shows clearly the way that components are connected for a given piece of equipment. It will use BS 3939 symbols to depict the components, and the layout and interconnections will bear no relation to the way that these will actually be on the equipment but will be drawn in the simplest and clearest way.

Wiring diagrams Wiring diagrams show how the components are to be wired and how and where the connections are to be made. The components are often drawn to physically look like what they represent and are often placed on the drawing in the positions that they will actually take up in the equipment.

Block diagrams Block diagrams, sometimes referred to as line diagrams, use square blocks to represent plant or equipment. Lines are drawn showing interconnections but these will not represent the actual wiring as in the two examples above. Details of what the blocks represent will either be written in the block or to one side. The interconnecting lines will have details written at the side indicating what cables these represent.

Examples of circuit, wiring and block diagrams are given on Information Sheet No. 1B.

Layout drawings An architectural plan of the building will have had the layout of plant and equipment added to it by the electrical designer as detailed above. The symbols representing this will be to BS 3939; these are different, however, to the ones used in circuit diagrams so care should be taken not to confuse the two.

Information Sheet No. 1B Circuit, block and wiring diagrams.

1. Circuit diagram

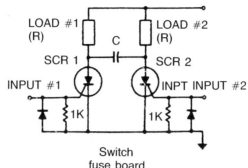

2. Block diagram

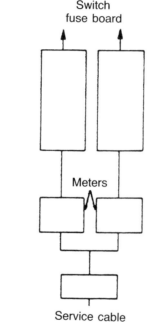

3. Wiring diagram

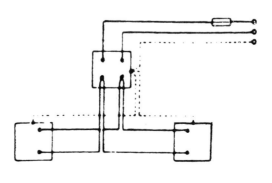

A representative selection of the most widely used BS 3939 symbols is shown on Information Sheet No. 1C and they are built up from certain basic shapes. For example a switch socket outlet with an indicating light will be depicted by the drawing of a socket outlet symbol with the symbol for a luminaire point added to it.

An example of a layout drawing is given on Information Sheet No. 1D.

Preparation of material lists

The electrician will often be asked to prepare a list of materials that are required for a particular section of the contract. These materials fall into the two main categories below.

Numbered items These are materials that can be counted, for example switch sockets, lighting switches or ceiling roses. The best way to approach this is to make yourself a 'Take Off Sheet'. This consists of a sheet of paper divided into columns for each of the items, which are indicated by drawing the symbol for each item at the head of each column (see Information Sheet No. 1E). The items are then counted for each section of the work and a total arrived at for each item.

Measured items These are materials such as cable, trunking or cable tray. These will require measuring either by visiting the site and physically measuring them, or by taking the measurements from the 'layout drawing'. If it is required to measure a length of a cable for example, a look at the drawing will show the route that the cable is taking. If the cable route is not shown, then this can be established by reading the specification and looking at the layout drawing. The drawing will be to scale so any measurement that is obtained will have to be multiplied by this. For example a measurement of 200 mm on a drawing with a scale of 1:100 would represent on site 200 × 100 or 20 000 mm, which is 20 m of cable.

Look at the drawing on Information Sheet No. 1D, it shows a layout drawing for a house extension. Using the BS 3939 on Information Sheet No. 1C, make yourself a Take Off Sheet and count the number of different items shown on the layout drawing. The quantities should match those on the Take Off Sheet on Information Sheet No. 1E. You must not forget that each accessory will require a backing box, and that the consumer unit will want protective devices and meter tails.

The scale of the drawing is 1:100 so 1 mm on the drawing will represent 100 mm on site. Measure the run of cable from the consumer unit to the cooker unit. This should be close to the measurement given on Information Sheet No. 1E. Do not forget to allow for the cable rising from the consumer unit to the ceiling void and also dropping down to the cooker unit at the other end. It is customary to allow an additional amount for termination of the cable at either end and in this case a total of 0.5 m was allowed for this.

Finally the materials should be written out clearly on a requisition sheet

Information Sheet No. 1C Electrical symbols: BS 3939.

Main control or intake point		Lamp or lighting point: wall mounted	
Distribution board or point *Note:* The circuits controlled by the distribution board may be shown by the addition of an appropriate qualifying symbol or reference		Emergency (safety) lighting point	
		Lighting point with built-in switch	
Main or sub-main switch		Lamp fed from variable voltage supply	
Socket outlet with interlocking switch		Projector or lamp with reflector	
Socket outlet with pilot lamp		Spotlight	
Multiple socket outlet *Example:* for 3 plugs		Single fluorescent lamp	
Socket outlet (mains): general symbol *Note:* In UK practice this general symbol normally implies the presence of an earthing contact. Exceptions to this rule should be indicated by a note, e.g. shaver outlet		Single-pole, one-way switch *Note:* Number of switches at one point may be indicated	
		Two-pole, one way switch	
		Three-pole, one way switch	
Switched socket outlet		Cord-operated single-pole one way switch	
Transformer		Two-way switch	
Consumer's earthing terminal	● E	Intermediate switch	
Electrical appliance: general symbol *Note:* If necessary use designations to specify type		Time switch	
		Switch with pilot lamp	
Fan		Push button	
Bell		Luminous push button	

12 Electrical Installation Practice 2

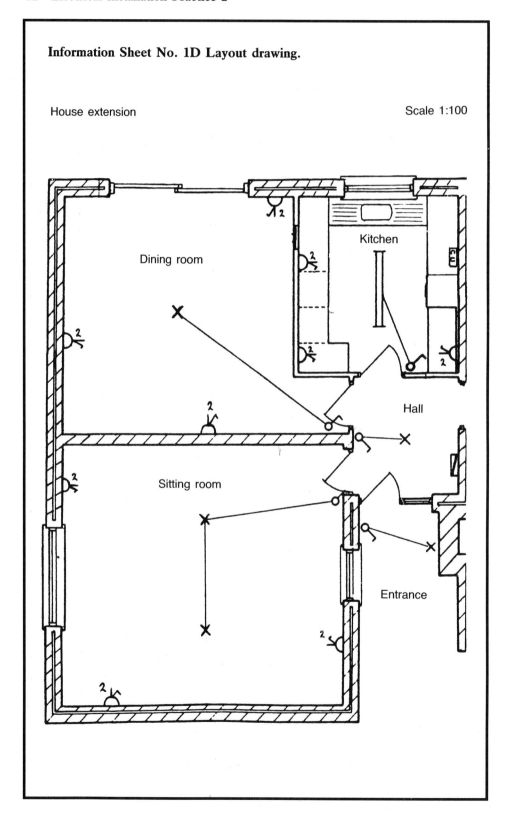

Information Sheet No. 1E Take off sheet.

AREA	²⚡	C.U.	⌀	×	☐	⚡	6mm²	2.5mm²	mm² 1.00	mm² 16		10m	70m	50m	2m																			
Extension	9	1	5	4	1	1						10m	70m	50m	2m																			
TOTALS	9	1	5	4	1	1						10m	70m	50m	2m																			

Information Sheet No. 1F Requisition sheet.

DEPARTMENT OF ROAD TRANSPORT AND MECHANICAL SERVICES

Workshop: _____ SITE _____

Materials Requisition

Goods required by -

Quantity	Description	Reference	Cost
9	2 Gang 13A Sw/Skt	M.K. 2747 WHI	
9	2 Gang K.O. Boxes	Appleby 665	
1	45A Flush Cooker Unit	Delta 2011	
1	Cooker Unit K.O. Box	Appleby 628	
5	1 Gang 1-Way Switch	M.K. 4870 WHI	
5	1 Gang P.D. Boxes	Appleby 623	
4	Ceiling Rose	Delta 1200	
4	B.C Lampholders	Delta 1000	
1	Twin Fluorescent	Thorn PP2675	
1	Diffuser	Thorn PPC6	
1	Bulkhead	Thorn 'Corsa'	
50 m	6 mm 6242Y cable		
100 m	2.5 mm 6242Y cable		
50 m	1.0 mm 6242Y cable		
1	Box Plastic Plugs		
1	Box 1 mm Clips	MM 8456	
1	Box 2.5 mm Clips	MM 8458	
1	Box 6 mm Clips	TWRK F6	
1	Box 1½" x 8 C.S.	Screws Nettle	
1	MEMRA3 6way C.U.	DEL 6QM	
6	DEL QCB M.C B.'s	DEL QCM	
2 m	16 mm^2 PVC.PVC. Singles	6181Y 16mm^2	
		TOTAL COST =	

Signed _J. Powell_____ DATE: _____

similar to the one shown on Information Sheet No. 1F. This is either forwarded to your company's office or taken directly to your stores. Some companies allow their senior site staff to order materials directly from the suppliers; in this case the details would be made out on a company's order form.

16 Electrical Installation Practice 2

Test 1

Choose which of the four answers is the correct one.

(1) In the construction of a building the most important person is the:

(a) Architect;
(b) Consultant;
(c) Client;
(d) Main contractor.

(2) The relationship between the client and the main contractor is:

(a) contractual;
(b) friendly;
(c) professional;
(d) positive.

(3) The symbols used to depict the position of electrical points on a layout drawing are to British Standard:

(a) BS 1361;
(b) BS 88;
(c) BS 1363;
(d) BS 3939.

(4) Company structures fall into two main types:

(a) horizontal and perpendicular;
(b) vertical and horizontal;
(c) vertical and perpendicular;
(d) primary and secondary.

(5) The Clerk of Works is responsible for the:

(a) design and layout drawings;
(b) measurement and finance;
(c) quality and workmanship;
(d) time keeping and wages.

Chapter 2
Electric Cables and Factors Determining Their Choice

2.1 Electric cables

Definition of an electric cable

In most cases, a cable may be defined as 'a length of insulated single conductor (solid or stranded), or two or more such conductors each provided with its own insulation which are laid up together...'. Thus a cable has two essential parts, a conductor and insulation.

Conductor materials The most widely used conductor material for electrical installations in buildings is copper. This is because copper has a low resistivity and is therefore a good conductor of electricity, it is easily drawn out into a wire and is comparatively cheap. Copper conductors are usually stranded to make them more flexible. Small cables designed for maximum flexibility have a large number of strands; they are called flexible cords (from 3 to 25 A current rating) and flexible cables (from 33 A upwards). Very small cable, commonly known as fittings or bell wire, is sometimes used for the internal wiring of luminaires and in bell circuits. In all stranded cables the strands are twisted together and the direction of the twist is called the 'lay'.

Another conductor material is aluminium; cheaper than copper, it is easier to handle in the bigger sizes because of its light weight. Aluminium, however, is not such a good conductor of electricity as copper, having a higher resistivity, and therefore a larger cross-sectional area (csa) cable is required for the same current-carrying capacity (see Information Sheet No. 2A). Special fluxes have to be used when soldering aluminium conductors and because of these reasons and the fact that it does not have the same mechanical strength as copper its use is confined to that of the larger size cables.

Insulation materials The materials chosen for the insulation are poor conductors, that is to say their resistivity is very high, so that they prevent leakage of electricity from the conductor. Insulating materials in common use at the present time include vulcanised butyl or silicone rubber, plastics such as polyvinyl chloride (PVC) and polythene, phenol-formaldehyde and magnesium oxide.

Cables are installed in a variety of different situations, and the electrician must take care that the the type of insulation on the chosen cable is suitable for that particular situation. Below are some of the working properties of the more common types:

- PVC – a good insulator, it is tough, flexible and cheap. It is easy to work with and easy to install. However, thermoplastic polymers do not stand

Information Sheet No. 2A Resistivity.

Resistivity is defined as 'the resistance of a material measured between opposite faces of a unit cube of that material'.

If, then, we were to take a cube of material whose side measured 1 cm (a cubic centimetre), and measured the resistance across it, we could use the result to determine the resistance of a sample of that same material whatever the length or cross-sectional area (csa).

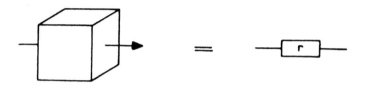

Now if the resistance of our unit cubic centimetre of material was to measure 1.5 ohms, then a ten centimetre length of the same csa would have a resistance of 15 ohms (resistance increases with an increase in length).

If we took that same ten centimetre length and doubled its csa then the resistance would be halved to 7.5 ohms (resistance decreases with an increase of csa).

The resistance of our sample was 1.5 ohms and this is called the 'resistivity' of that material. If we were to measure the resistance of unit cubes of other materials, each would have a different result because the 'resistivity' of materials is different.

We can say then that the resistance of a material depends upon its length, its csa and the resistivity of the material. This resistivity is given the greek letter 'rho' (ρ) as a symbol and the formula used to calculate the resistance of a material is:

$$R = \frac{\rho \times l}{a}$$

where
R = Resistance
ρ = Resistivity
l = length
a = csa.

up to extremes of heat or cold, and the IEE Wiring Regulations recommend that ordinary PVC cables are not used in temperatures above 60°C and below 0°C if those sort of temperatures are going to persist. Care should be taken with PVC cables when using blowlamps or burning off old cable as this form of insulation gives off toxic fumes when burnt.
- synthetic rubbers – such as vulcanised butyl, withstand high temperatures much better than PVC and are therefore much used for the connection of such things as immersion heaters, storage heaters and boiler house equipment.
- silicon rubber – FP 200 cable using silicon rubber extruded over aluminium foil is becoming more popular in recent times for the installation of such things as fire alarms. This is due largely to the fact that it retains its insulation properties after being burned and is somewhat cheaper than mineral insulated metal sheathed (MIMS) cables.
- magnesium oxide – is the white powdered substance used as insulation in mineral insulated cables. This form of insulation is hygroscopic in nature and therefore must be protected from damp by the use of special seals. Mineral insulated cables are able to withstand very high temperatures indeed and being metal sheathed are able to stand up to a high degree of mechanical damage too.
- phenol-formaldehyde – is a thermosetting polymer much used in the production of such things as socket outlets, plug tops switches and consumer units and is able to withstand temperature in excess of 100°C.

2.2 Factors determining choice of cable

Environmental factors

We have seen that electric installations must be suitable for the purpose that they have been installed, this makes the choice of type of cable to be used of extreme importance. Factors determining the choice of cables are:

- Ambient temperature;
- Moisture present;
- Danger of electrolytic action;
- Proximity to corrosive substances;
- Possible damage by animals;
- Exposure to direct sunlight;
- Possibility of mechanical stress (suspension);
- Mechanical damage.

Ambient temperature Current-carrying cables produce heat and the rate that the heat is able to be dissipated depends upon the temperature surrounding the cable. If the cable is in a cold situation then the temperature difference is greater and there can be a substantial heat loss. Should the cable be in a hot situation then the temperature difference of the cable and its surrounding environment will be small and little if any of the heat will be dissipated. Problem areas are boiler houses and plant rooms, thermally insulated walls and roof spaces, and production plants using great heat such as steel and glass works. Appendix 9 of the IEE Wiring Regulations gives tables of different types of cable and it will be seen that the current rating changes in accordance with the conditions under which the cable is installed.

Moisture Water and electricity do not mix and care should be taken at all times to avoid the ingress of moisture into any part of an electrical installation by the use of watertight enclosures where appropriate. Any cable with an outer PVC sheath will resist the penetration of moisture and will not be affected by rot; however, suitable watertight glands should be used for the termination of these cables.

Electrolytic action Dissimilar metals together with the presence of moisture can cause electrolytic action resulting in the deterioration of the metal and care should be taken to avoid this. An example of this is when brass glands are used with galvanised boxes in the presence of moisture. Metal sheathed cables can suffer when run across galvanised sheet structures, and if aluminium cables are to be terminated on to copper bus-bars then the bars should be tinned.

Proximity to corrosive substances The metal sheaths, armour, glands and fixings of cables can all suffer from corrosion when in close proximity to certain substances. Some of these are the magnesium chloride used in the construction of floors, plaster undercoats containing corrosive salts, unpainted walls of lime or cement, oak and other types of acidic wood. Metal work should be plated or given a protective covering.

Possible damage by animals Cables installed in situations where rodents are prevalent should be given additional protection or installed in conduit or trunking as these animals will gnaw cables and leave them in a dangerous condition. Installations in farm buildings should receive similar consideration and should if at all possible be placed well out of reach of the animals to avoid the effects of rubbing, gnawing and urine.

Exposure to direct sunlight Cables sheathed in PVC should not be installed in positions where they are exposed to direct sunlight because this causes them to harden and crack. The reason for this is that the ultraviolet rays leach out the plasticiser in the PVC leaving it hard.

Mechanical stress PVC cables when used for overhead wiring between buildings can be subject to mechanical stress if a catenary wire is not used to support them along the way (see Information Sheet No. 2B). Flexible cables used to suspend heavy luminaires which exceed the recommended weights given in IEE Regulation 523−32 will feel the effects of mechanical stress. These cables can suffer from stress too when subjected to excessive vibration causing breakdown of the insulation.

Mechanical damage All conductors and cables should be guarded from damage by the provision of additional protection where they pass through

Electric Cables and Factors Determining Their Choice 21

Information Sheet No. 2B Suspension of overhead cables.

1. Road crossings accessible to vehicles

All methods of suspension
5.8 m minimum above ground

2. Accessible to vehicles but not a road crossing

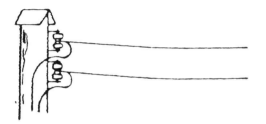

All methods of suspension
5.2 m minimum above ground

3. Inaccessible to vehicles

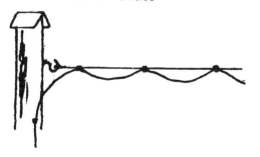

PVC cables supported by a catenary wire.
3.5 m minimum above ground

For further details see Table 11 B of the IEE Wiring Regulations.

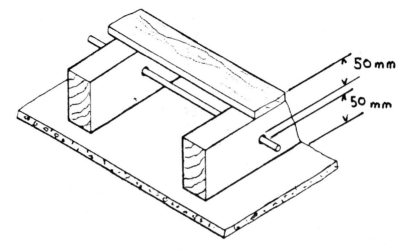

Fig. 2.1 Cables run in floor joists.

floors and walls or are installed in exposed positions where damage could occur. Cables to be installed underground should have incorporated into them a sheath or armouring resistant to any mechanical damage likely to occur. When cables pass through holes drilled in wooden joists these should be 50 mm from the top or bottom of the joist measured vertically (see Fig. 2.1). Numerous other examples of situations where cables can be subjected to mechanical damage can be found in section 523–19 to 523–33 of the IEE Regulations.

2.3 The selection of conductor sizes

Factors determining selection

Once a decision as to the type of cable suitable for the environmental conditions of the installation has been made, then the task of determining the size of the conductor to be used must be undertaken. The factors to be taken into consideration are:

- Design current of the circuit and rating of protective device;
- Control of voltage drop;
- Regulation of thermal constraints;
- Shock protection.

How the above criteria are taken into consideration when calculating the size of the conductor is best demonstrated by following a worked example.

A fuse to BS 3036 is to feed a resistive load comprising of a 3 kW immersion heater. The cable which is of the twin with earth general purpose PVC insulated PVC sheathed type, is to be clipped directly to the side of the wooden floor joists. The length of run is 18 m, it will be bunched with two other cables and it is estimated that the ambient temperature will be 35°C at its worst.

Design current of the circuit For DC loads or AC resistive loads the load current can be worked out by the following formula:

$$I = \frac{P}{V} = \frac{3000}{240} = 12.5 \text{ A}$$

If no 'diversity' is to be allowed for in accordance with Tables 4A and 4B of Appendix 4 of the Regulations then the load current will be the design current (I_b).

Selection of protective device Having established the design current of the circuit I_b, the next stage is to select the overcurrent device of the correct type and rating. We have been told that the type of protection is a fuse to BS 3036 which is a semi-enclosed rewirable fuse and it can be seen from Table 53A of the Regulations that the nearest sizes are a plain or tinned copper wire of nominal diameter 0.35 mm with a nominal current of 10 A, or a 0.5 mm with a nominal current of 15 A. This rating will be the nominal circuit current (I_n); obviously, as stated in Appendix 9 Section 5 of the Regulations, this must not be less than the design current of the circuit I_b, and the 15A fuse is selected.

Voltage drop Having established the design current I_b of the circuit under consideration, the next step is to identify the length of the cable run and the permissible voltage drop for the equipment being supplied, as this may be an overriding consideration.

The permissible voltage drop in mV, divided by I_b and the length of run will give the value of volt drop in mV/A/m which can be tolerated. A volt drop not exceeding that value is identified in the appropriate table and the corresponding csa of conductor needed on this account can be read off directly before any other calculations are made as follows. Regulation 522–8 states that the volt drop from the origin of the circuit shall not exceed 2.5% of the nominal voltage, which in this case is 240 V.

$$\text{Maximum voltage drop allowed} = \frac{240 \times 2.5}{100}$$
$$= 6\text{V}$$

$$\text{Maximum mV/A/m} = \frac{\text{mV}}{I_b \times \text{m}}$$
$$= \frac{6000}{12.5 \times 18}$$
$$= 26.66 \text{ mV/A/m}$$

By referring to Table 9D2, any cable with 26.66 mV/A/m or less will therefore give an actual volt drop of 6V or less.
From Table 9D2:

$$2.5 \text{ mm}^2 = 18 \text{ mV/A/m (hence volt drop less than 6V)}$$

i.e.
$$\frac{18 \times 12.5 \times 18}{1000} = 4.05 \text{ V}$$

This is below the 6V maximum allowable volt drop so 2.5 mm² cable is selected.

Note: The above cable selection shows compliance only with Regulation 522–8 and item 7 of Appendix 9.

Correction factors The tabulated current-carrying capacity (I_b) and voltage drop data in the IEE current rating tables may be applied as the relevant data under the following basic conditions.

(1) Frequency is 50 to 60 Hertz ± 2%.
(2) Ambient temperature is 30°C.
(3) Conductors identical equally and fully loaded.
(4) Grouping is limited to one multi-core cable run separately, or one circuit of single core cables bunched together. Alternatively clearance between cables is two cable diameters or more.
(5) The cable type is appropriate for the method of installation and its environment.
(6) Maximum conductor operating temperature is not exceeded (see 3.11)
(7) Power factor is not worse than 0.6 for conductors up to 120 mm² or 0.8 for larger conductors (lagging).
(8) The overcurrent protective devices are HBC fuses to BS 88 or BS 1361. Alternatively a circuit breaker to BS 4752 Part 1 or an MCB to BS 3871 could be used.

Where any of the above conditions differ, or when the cable is in contact with thermal insulation, or where the overcurrent protection device is a semi-enclosed rewirable fuse, then an appropriate correction factor must be applied for each of the conditions.

A definition of these, together with symbols where relevant, is to be found in section 4 of Appendix 9 of the IEE Wiring Regulations.

Correction factors can be found in the IEE Regulations as follows:

Ambient temperature	C_a	Table 9C1 and 2
Grouping factors	C_g	Table 9B
Thermal insulation	C_i	Reg. 522–6
Operating temperature	C_t	App. 9, Sec. 7.1
BS 3036 Fuse		App. 9, Sec. 5

Application of correction factors How do the above conditions affect the cable in our example?.

If the conditions in which the cable was installed were as the basic conditions laid out above, then the tabulated current-carrying capacity and voltage drop data stated above could be applied as the relevant data and we could use the 2.5 mm² cable. If, however, any of the factors mentioned are applicable then they must be applied, and the conductor size necessary for continuous current-carrying capacity (I_z) and for overload determined.

A look at our example shows that factors affecting the cable are ambient temperature (C_a), the grouping of cables (C_g) and the factor for using BS 3036 fuses.

Whilst all correction factors affecting I_z (that is C_a, C_g, and C_i) can, if desired, be applied to the values of tabulated current (I_t) as multipliers, giving the effective I_z for the installation conditions concerned, this involves a process of trial and error until a csa is reached which insures that I_z is not less than I_t and not less than I_n of any protective device that is selected. In any event, if a correction factor for protection by semi-enclosed fuse is necessary, this has to be applied to I_n as a divisor. It is therefore more convenient to apply all the correction factors to I_n as divisors.

How is this applied to the cable in our example?.

We have already agreed that the fuse would be the 15 A one giving us a nominal circuit current I_n of 15 A.

From the IEE Wiring Regulations:

$$C_a \text{ (ambient temperature)} = 0.97$$
$$C_g \text{ (grouping factor)} = 0.70$$
$$\text{BS 3036 fuse} = 0.725$$

Therefore:

The value of tabulated current required
$$= \frac{I_n}{C_a \times C_g \times \text{BS 3036}}$$
$$= \frac{15}{0.97 \times 0.70 \times 0.725}$$
$$= 30.61 \text{ A}$$

From Table 9D2 Column 6

$$\text{Cable selected} = 4 \text{ mm}^2$$

It should be noted that when the cross-sectional area is selected in accordance with the recommendations of Appendix 9 Section 6, which is the method used above, the value of I_t chosen is not I_z. It is not necessary to know I_z when this method is used. If it is desirable to know I_z, then the first method mentioned, of using the factors as multipliers to I_t can be adopted.

Test 2

Choose which of the four answers is the correct one.

(1) The most widely used conductor material for cables is:

(a) magnesium oxide;
(b) phenol-formaldehyde;
(c) copper;
(d) lead.

(2) The initials PVC stand for:

(a) poly vinyl chlorine;
(b) polyvinyl chloride;
(c) plastic vinyl coating;
(d) plastic vinyl chlorine.

(3) The ambient temperature surrounding a cable is the:

(a) lowest working temperature;
(b) highest working temperature;
(c) normal working temperature;
(d) absolute working temperature.

(4) The tabulated current-carrying capacity of cables is the:

(a) current rating given in the IEE Wiring Regulation Tables;
(b) current rating after applying correction factors;
(c) maximum demand of the circuit;
(d) assumed current demand of the circuit.

(5) Magnesium oxide is said to be:

(a) hydroscopic in nature;
(b) water resistant;
(c) an excellent conductor;
(d) hygroscopic in nature.

Chapter 3
Protection Against Indirect Contact

3.1 Earthing

The general mass of earth

The expression 'EARTH' is defined in Part 2 of the IEE Wiring Regulations as:

> The conductive mass of the Earth, whose electric potential at any point is conventionally taken as zero.

The 'Electricity Supply Regulations 1988' require the local supply authority to solidly connect to earth one conductor of their AC distribution network. This is connected at the secondary side of the supply transformer (star point) to an earth electrode which is buried in the 'general mass of earth' (see Information Sheet No. 3A).

From Book 1 we are aware that if we were in contact with the general mass of earth and we were to accidentally make 'direct contact' with a current-carrying conductor, or make 'indirect contact' by touching metalwork of the installation that had become live due to a fault, we would receive an electric shock. We learnt that the current would pass through our bodies, through the 'general mass of earth' and return to the star point of the supply authority's transformer so completing the circuit. What can we do to protect ourselves and others from this?

Protection against electric shock

We can protect ourselves and others from coming into direct contact with current-carrying conductors by the use of insulation, barriers, enclosures or by placing out of reach (see Regulation 412–1), but how can we protect ourselves against receiving an electric shock from an 'exposed conductive part', such as the metalwork associated with our installation, that has become live due to a fault? We could use Class II Equipment, i.e. equipment that does not rely on the basic insulation alone but has supplementary insulation provided. Examples of these are all insulated electric drills, hair driers etc. which have no provision for connection of a protective conductor but rely on the additional insulation for the protection of the user. Alternatively we could achieve protection against indirect contact by the use of electrical separation, which is the use of isolating transformers such as those used in electric shaver sockets to BS 3052 for example (see Regulation 413–

28 Electrical Installation Practice 2

Information Sheet No. 3A Earthing.

1. Connections to general mass of earth

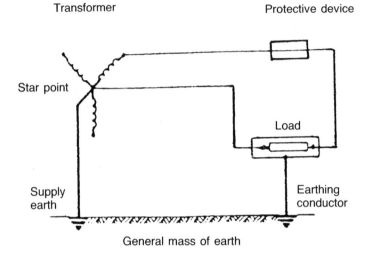

2. Earth fault loop path

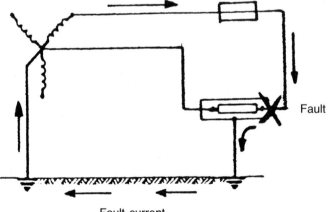

Protection Against Indirect Contact

1). We could even go to the trouble of placing our equipment in a non-conducting location (see Regulation 413–27 to 31). While these methods achieve the necessary protection required for particular sets of circumstances they are by no means practical for the greater part of electrical installations. We therefore must look to another alternative suggested by Regulation No. 413–2; that of 'equipotential bonding'.

3.2 Equipotential bonding

The reasons for it

By connecting all the exposed conductive parts of the installation and any other potentially dangerous metalwork (extraneous conductive parts) together and connecting these to the general mass of earth, it can be ensured that everything is at the same potential as earth and that:

(1) no dangerous differences in potential can exist between the metal work of different parts of the installation and between the metal work and earth;
(2) a path is provided for earth leakage currents which can be detected and interrupted by a suitable protective device (more on this later).

Main equipotential bonding

In each electrical installation a consumer's main earthing terminal bar will be provided in accordance with Regulation 542–1. Main equipotential bonding conductors shall connect to the main earthing terminal of the installation the following:

- Main water pipes;
- Gas installation pipes;
- Other services pipes and ducting;
- Risers of central heating and air conditioning systems;
- Exposed metallic parts of the building structure.

Connection to services may require the permission of the undertakings responsible, although compliance with Regulation 413–2 will normally satisfy the requirements of the Secretary of State for Energy regarding Protective Multiple Earthing (PME).

Equipotential bonding to gas and water services should be made as soon as practical after their entry into the building, connections being made on the consumer's side of the meter (within 600 mm) in the case of gas, and the installation side of any insulating section or insert (stopcock) where fitted, for the water (see Information Sheet No. 3B).

The size of main equipotential bonding conductors

The cross-sectional area (csa) of the main equipotential bonding conductors shall be not less than half the csa of the earthing conductor of the installation,

Information Sheet No. 3B Bonding connections.

1. Main equipotential bonding

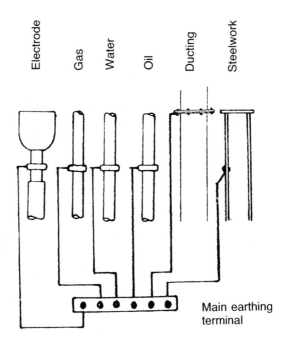

2. Supplementary bonding

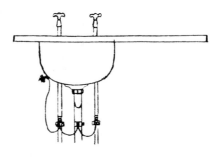

subject to a minimum size of 6 mm^2 copper cable. Where PME (TN−C−S) conditions apply, the maximum csa need only be 25 mm^2 copper cable, though the local supply authority should be consulted for any special requirements in your area. Aluminium or copper clad cables are not recommended for connection to pipes where condensation might be present during normal use.

Equipotential bonding zone

This main equipotential bonding is intended to create a zone in which any voltages between exposed conductive parts and extraneous conductive parts are minimised (Regulation 413−2, Note 1). However it could well be that within the zone formed by the main equipotential conductors, local 'supplementary bonding' connections will need to be made to metal parts, to maintain the equipotential bonding zone (see Regulation 413−7).

Equipment outside the equipotential bonding zone

Sockets serving portable equipment that is to be used outside the equipotential bonding zone, for example those in garages or workshops, are required to be protected by a residual current device with an operating current not exceeding 30 mA. The socket will carry a label saying 'FOR EQUIPMENT OUTDOORS', see Regulation 471−12 and 47. Where fixed equipment outside the equipotential zone is supplied from a source inside the zone, and that equipment can be touched by a person in contact with the general mass of earth, then the earth fault loop impedance shall be such that disconnection occurs within 0.4 s.

3.3 Supplementary bonding

Why it is required

Supplementary bonding conductors 'may' be required in situations such as kitchens where a person may be in contact with an extraneous conductive part such as a metal sink, and an exposed conductive part such as the metalwork of a washing machine. Taps often have an insulated washer between them and the metal sink and some plumbing fittings have insulating material between them and the pipework. In cases such as this, supplementary bonding or a combination of this and permanent reliable extraneous conductive parts may be required to ensure that all the metal work is at 'earth potential' (see Regulation 547−7). An example of supplementary bonding can be found on Information Sheet No. 3B.

In a room containing a fixed bath or shower, supplementary bonding 'shall' be provided between accessible exposed conductive parts of equipment such as instantaneous shower units, immersion heaters and extract fans that can be touched simultaneously. It is also required between the exposed conductive parts previously mentioned and accessible extraneous conductive parts such as radiators, taps, pipes etc. that can be touched simultaneously, and also between accessible extraneous conductive parts that can be touched simultaneously (see Regulation 471−35).

The size of supplementary bonding conductors

The csa of a supplementary bonding conductor connecting two exposed conductive parts shall not be less than that of the smaller protective conductor connected to the exposed conductive parts subject to a minimum of 2.5 mm^2 if mechanically protected and 4 mm^2 if not (see Regulation 547−5).

A supplementary bonding conductor connecting exposed conductive parts to extraneous conductive parts shall have a csa of not less than half that of the protective conductor connected to the exposed conductive part, subject to a minimum of 2.5 mm^2 if mechanically protected or 4 mm^2 if not (see Regulation 547−6).

The csa of a supplementary conductor connecting two extraneous conductive parts shall be 2.5 mm^2 if mechanically protected or 4 mm^2 if not, except that if one extraneous conductive part is also connected to an exposed conductive part in compliance with Regulation 547−5, that regulation shall apply to the conductor connecting the two extraneous conducting parts (see Regulation 547−6).

3.4 Circuit protective conductors

Definition and use

The IEE Wiring Regulations define a protective conductor as a conductor used for some measure of protection against electric shock. The abovementioned cables can all be considered to be 'protective conductors' under these definitions.

A 'circuit protective conductor' (cpc), which used to be known as an earth continuity conductor, is defined as a conductor connecting exposed conductive parts of equipment to the main earthing terminal. The cpc can be one of the following:

(1) part of a composite cable (Twin and earth 6242Y) where it will come under the requirements of Chapter 52 of the Regulations;
(2) a single core cable (single insulated 6491X) colour coded with green and yellow stripes, or bare conductor subject to Regulation 542−16, Regulation 543−1 and 1a, Regulation 543−15 and 15a;
(3) the metal sheath of a cable and comply with Regulation 543−7 items (i) and (ii);
(4) a metallic conduit or other metal enclosure which will satisfy Regulation 543−9 to 19.

Particular cpcs, such as those forming part of a ring final circuit for example, will be dealt with in the appropriate sections of the book.

Protection Against Indirect Contact

The size of circuit protective conductors

The csa of every protective conductor other than the equipotential bonding conductors mentioned above shall be determined by:

(1) calculation according to Regulation 543−2, or
(2) selection in accordance with Regulation 543−3.

It will be seen that Regulation 543−3 says that reference should be made to Table 54F. A look at this table will show that for cables 16 mm^2 and below, the cpc should be the same size as the phase and neutral conductors. This would mean that all the twin and earth 6242Y cables mentioned above would be contrary to the Regulations because they all have cpcs smaller than their respective phase and neutral conductors, with the exception of 1 mm^2 which has the same size cpc.

Clearly it was not the intention that composite cables should have their cpcs increased in accordance with the table and the calculation in (1) above should be applied (see Appendix 8).

From the designer's point of view it is often advantageous to use the calculation method illustrated in Regulation 543−2 as savings can often be made. The method of using the tables to calculate the csa of cpcs is shown in Information Sheet No. 3C.

3.5 Connecting protective conductors and earth electrodes

Earth clips

The major part of the equipotential bonding that takes place on an installation takes the form of connections to water pipes. This is achieved by the use of specially designed earth clips and it is important that the screw which secures the cable to the clip is not the same one that tightens the clip to the pipe. The pipe work is cleaned by the use of steel wool or emery paper, and the strap passed round the pipe and through the clip. It passes under a screw which, once the strap has been tightened as much as possible by hand, is screwed home with a suitably sized screwdriver. There is a locking nut provided and this must now be tightened to avoid the screw coming loose due to vibration. A warning notice that says 'SAFETY ELECTRICAL CONNECTION DO NOT REMOVE' is slipped onto the cable which is then stripped and secured under the clamp (see Information Sheet No. 3D).

Earth electrodes

A number of different types of earth electrodes are recognised for the purposes of the IEE Wiring Regulations and are detailed in Regulation 542−10 as follows:

- Earthrods or pipes;
- Earth tapes or wires;

Information Sheet 3C. Regulation 543–2.

IEE Wiring Regulation No. 543–2 states that where the csa of protective conductors are to be calculated then they shall not be less than the value obtained by the use of the following formula. It is applicable to disconnection times of 5 s or less.

$$S = \frac{\sqrt{I^2 t}}{k} \text{ mm}^2$$

where

S is the csa in mm^2;
I is the fault current in amperes;
k is a factor depending on how high the temperature of the cable can safely be allowed to rise

The following steps are necessary to establish the csa of the cpc:

(a) Determine the maximum permitted fault loop impedance Z_s from Table 41A1 or 41A2.
(b) I can be found by applying the formula:

$$I = \frac{V}{Z_s}$$

(c) Find out the maximum circuit disconnection time from Appendix 8, Fig 8 to 15.
(d) Determine the value of k from Table 54B, C, D or E.
(e) Apply the formula given in Reg. 543–2.

Protection Against Indirect Contact

Information Sheet No. 3D Connecting protective conductors.

1. Earth electrode

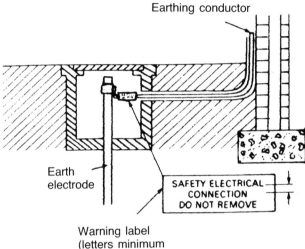

Warning label
(letters minimum
4.75 mm high)

2. Earth clip

- Earth plates;
- Earth electrodes embedded in concrete;
- Metallic reinforcement of concrete;
- Metallic pipe systems where not precluded by Regulation 542–14 and 15;
- Metallic cable sheaths where not precluded by Regulation 542–15;
- Other suitable underground structure.

Connection of the earthing lead is made by a substantial clamp to the earth electrode and a warning label fitted. It is important that the electrode is accessible for regular inspection and testing and this is usually achieved by the fitting of a concrete inspection cover over the head of the electrode (see Information Sheet No. 3D). The electrode must be protected against mechanical damage, corrosion, the soil drying out and freezing of the ground for which due allowance must be made (see Regulation 542–11 to 18).

3.6 Protection by automatic disconnection from the supply

It was seen in Section 3.2 above that the bonding formed a path for leakage currents which could be detected and would lead to the interruption of the supply by a suitable protective device.

Regulation 413–3 states that the following items must be co-ordinated so that during an earth fault the voltages between simultaneously accessible exposed and extraneous parts occurring anywhere in the installation will be of such a size and length of time as not to cause danger:

- The earthing arrangements of the installation;
- The relevant impedance of the circuits concerned; and
- The characteristics of the protective device.

The earthing arrangements were dealt with in Section 3.2 above, so let's take a closer look at the second of the items, the impedance of the circuit.

Earth fault loop impedance

It can be seen from the sketch in Information Sheet No. 3A that the path which the fault current takes is called the 'earth fault loop path'. The type of soil has a bearing on how effectively the fault current flows through the general mass of earth, clay being the best and sandy soil being poor. If the soil is really dry this will further 'impede' the flow of fault current and affect the operating time of the overcurrent protective device which is determined by this current. It is imperative, therefore, in the interest of the safety of the consumer that an 'impedance test' is carried out to ensure that the impedance of the earth fault loop is below the required limit (see Chapter 10 of this book).

The maximum values for earth fault loop impedance are given in Tables 41A1 and 41A2 of the regulations and depend on the the type and rating of the protective device.

Protection Against Indirect Contact

Co-ordination of earthing systems and protective devices

The final part of the installation which requires co-ordinating to ensure disconnection of the supply should a fault occur, is the protective device; this was discussed briefly in Chapter 5 of Book 1. How these protective devices are co-ordinated with the earthing system provided is as follows:

TN-S system All exposed conductive parts of the installation will be connected by protective conductors to the main earthing terminal of the installation and that in turn will be connected to the earth point of the supply authority. The 'protective devices' shall be either overcurrent devices or residual current devices (see Regulation 413−8).

TT system Exposed conductive parts shall be connected by protective conductors individually, in groups or collectively to an earth electrode or electrodes. The 'protective devices' shall be either overcurrent or residual current devices; however, residual current devices are preferred (see 413−10 and 12). Where protection against indirect contact is by means of equipotential bonding in a TT system, then every socket outlet shall be protected by a residual current device (see Regulation 471−13).

TN-C-S system All exposed conductive parts of the installation will be connected by protective conductors to the main earthing terminal of the installation. The earthing lead provided by the consumer will be connected by the supply authority to the neutral or PEN conductor of the supply (PME system). The 'protective device' will be of the overcurrent type; residual current devices are not to be used where the neutral and protective functions are combined in one conductor (see 413−9).

As discussed in Book 1 there are two other earthing systems in existence. The IT system which the Supply Regulations do not allow to be used in this country and the TN-C system which can only be used when special authorisation has been obtained or private plant is utilised. Further details of these systems can be obtained if required from Appendix 3, Regulation 413−13 to 17 and Regulation 542−5.

As indicated in item (i) of Regulation 413−4 other methods of complying with Regulation 413−3 (co-ordination) are not precluded and an alternative method of complying with Regulation 413−3 for circuits supplying sockets is explained in Appendix 7 of the IEE Wiring Regulations.

3.7 Residual current devices

The residual current device (RCD) is connected as shown in Fig. 3.1. It will be noted that the phase and neutral currents pass through identical coils wound in opposite directions on a toroid or laminated circular metal ring. A third winding or detector coil is connected to the trip coil of the main contact breaker as shown. During normal operation of the circuit the fluxes created by the phase and neutral windings are opposite and equal and no

38 Electrical Installation Practice 2

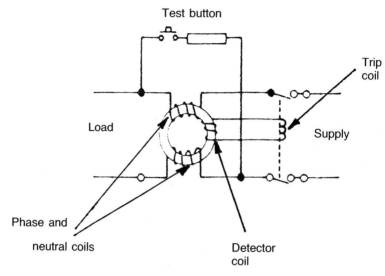

Fig. 3.1 Residual current device

voltage is detected by the detector coil. Should either of the currents flowing through the phase and neutral coils vary due to an earth fault in the circuit, then the fluxes will be out of balance. This will result in a magnetic flux being set up in the toroid which induces an e.m.f. in the detector coil. This results in a current in the trip coil which operates the main contactor and the circuit is disconnected from the supply.

A RCD is suitable for protection against indirect contact provided it will operate at 30 mA. If the fault loop impedance is high this could lead to dangerously high voltages in the earthed metalwork, so the regulations prohibit the use of these devices in TN and TT systems where the product of the rated residual operating current in amperes and the earth fault loop impedance in ohms is greater than 50 V (see Regulation 413–6).

Overcurrent devices are discussed in the following chapter.

Test 3

Choose which of the four answers is the correct one.

(1) The initials cpc stand for:

(a) circuit protective conduit;
(b) circuit protection conductor;
(c) connecting protective conductor;
(d) circuit protective conductor.

(2) The final connection to an earth electrode will carry a label saying:

(a) Danger Electrical Connection Do Not Remove;
(b) Electrical Connection Danger Do Not Remove;
(c) Safety Electrical Connection Do Not Remove;
(d) Electrical Safety Connection Do Not Remove.

(3) Contact by a person with exposed conductive parts made live by a fault is described as:

(a) direct contact;
(b) indirect contact;
(c) insufficient contact;
(d) sufficient contact.

(4) The path which the fault current takes through the general mass of earth to the star point of the supply transformer is:

(a) general mass of earth path;
(b) earth fault current route;
(c) fault current route path;
(d) earth fault loop path.

(5) Sockets serving portable equipment to be used outside the equipotential bonding zone shall have a label stating:

(a) 'FOR EQUIPMENT OUTDOORS';
(b) 'DANGER HIGH VOLTAGE';
(c) 'RCD PROTECTED';
(d) 'OUTDOOR EQUIPMENT'.

Chapter 4
Control and Protection for the Consumer

4.1 Consumer's switchgear

We saw in Book 1 how the supply authority brought their cables from the local sub-station into the consumer's premises and terminated it at the 'service cut-out'. In the case of the larger premises this would be a three phase and neutral (TP&N) four wire supply and for the smaller premises a single phase and neutral (SP&N) two wire supply. The equipment contained provision for the automatic disconnection of the supply under short circuit, overload and earth fault conditions. From the service cut-out the supply authority's cables went to the electric meter, before connection was made to the consumer's switchgear. We learnt how for the domestic and smaller premises a 'consumer unit' to BS 1454 complied with the requirements of the IEE Regulations for the control of electrical installations and details were given for the installation of this. The sequence of control equipment is shown in Fig. 4.1.

Larger premises are on a TP&N supply, to help spread the load over the three phases of the supply authority's distribution, and to accommodate the three phase equipment and machinery that may be installed in this type of building. It will be necessary therefore to install a TP&N distribution board which will have a bank of fuse-ways for each of the three phases. This will give the additional fuse-ways required for the larger installation and give the facility of a three phase 415 Volt supply. Because the distribution board does not have an internal switch like that of the consumer unit, a separate

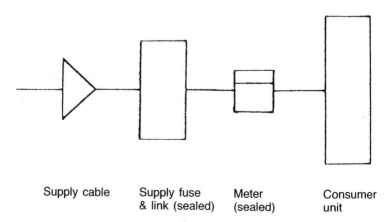

Fig. 4.1 Sequence of control equipment

Control and Protection for the Consumer 41

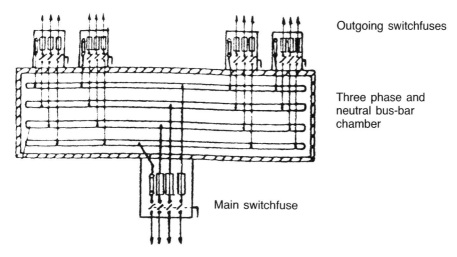

Fig. 4.2 Bus-bar interconnections

switchfuse will have to be provided for control of the distribution board to comply with Regulation 476–15. A representative selection of switchfuses and distribution boards are shown on Information Sheet No. 4A.

For industrial premises it is quite unlikely that one distribution board will be sufficient for the purpose, therefore several will have to be installed. This is achieved by the fitting of a bus-bar chamber (see Information Sheet No. 4A) to a fused switch and the individual switchfuses controlling the individual distribution boards fitted on top of this. The interconnecting cables go to each of the phases and neutral in turn (see Fig. 4.2).

An alternative method of control to the above is the fitting of a:

- 'Sub-distribution board'. The fused switch is installed as before, but this time it connects to a heavy duty TP&N;
- Distribution board. This is installed either adjacent to the mains intake position, or at a suitable remote position to suit the distribution of the sub-mains cables to distribution boards in other parts of the building. If the sub-distribution board or any of the other distribution boards are situated in a detached building, then an additional means of isolation will be provided at that point also (see Fig. 4.3).

Where separate tariffs are in operation, for example a school kitchen which will have a different tariff for energy used for catering to that of its lighting and power, or an 'off peak' supply installation using separate meters, then each should be treated as a separate installation. This will mean the installation of separate meters by the supply authority and separate switchgear by the consumer, see Book 1. An off peak supply using a 'white meter', however, will have a normal installation, the different tariffs being achieved by a timer and relay automatically switching over from one tariff to the other inside the meter. There are a number of variations on the above, but whichever method is chosen the installation of the switchgear will have to comply with the IEE Wiring Regulations so let us look at some of these now.

Information Sheet No. 4A Switchgear.

1. Switchfuses

2. Bus-bar chamber

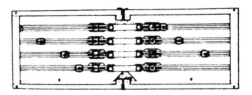

3. Distribution boards

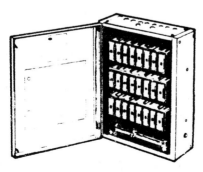 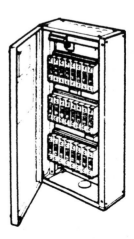

Control and Protection for the Consumer 43

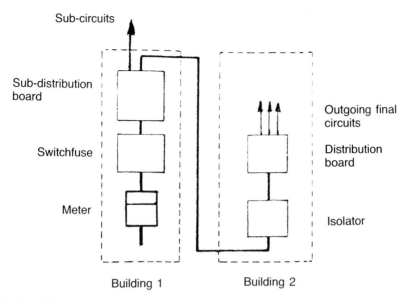

Fig. 4.3 Sub-distribution board in a detached building

4.2 IEE Regulations concerning switchgear

Every electrical installation must be provided with a means of isolation and be protected against short circuits, overloads and dangerous earth fault currents. In addition, means for switching off for mechanical maintenance, emergency switching or both shall be provided. Very often the functions of switching and protection etc. are incorporated in the same device, i.e. a switch fuse (see Regulation 476–1).

The means provided for non-automatic isolation and switching shall isolate each of the live supply conductors. However, the following shall not incorporate a means of isolation or switching:

- The PEN conductor of the TN–C system;
- The protective conductors in the TN–S, TN–C–S and TT systems;
- The neutral in a TP&N supply (except as Chapter 55 in IEE Regs.).

Multi-pole switches must be linked so that the neutral never breaks before the phase conductors and never makes after them.

The positions for 'on' and 'off' should be clearly indicated and indicate only when the switching operation is complete. The equipment should be designed in such a way that inadvertent switching 'on' of the supply cannot occur.

If the purpose of the switchgear is not immediately obvious, then a label must be attached indicating that purpose. Labels indicating the presence of voltages in excess of 250 V should be attached to switchgear and associated enclosures, and if two single phase items of equipment, connected to different phases of the supply, can be touched simultaneously, a notice indicating the level of voltage shall be fixed to them.

All conductors and protective conductors should be clearly identified and diagrams and charts supplied showing the current rating and purpose of each fuse or circuit breaker.

Where an installation uses a RCD a notice shall be fixed in a prominent position on or near the main distribution board. The notice shall be in indelible characters not smaller than 11-point (2.5 mm) high and shall say the following:

> This installation, or part of it, is protected by a device which automatically switches off the supply if an earth fault develops. Test quarterly by pressing the button marked 'T' or 'TEST'. The device should switch off the supply and should then be switched on to restore the supply. If the device does not switch off the supply when the button is pressed, inform your electrical contractor.

A notice shall be fixed on or near the Main Distribution Board of 'every' installation and shall be of a durable material that will ensure legibility throughout its life. The notice shall be in indelible characters not smaller than 11-point (2.5 mm) high and shall say the following:

IMPORTANT
> This installation shall be periodically inspected and tested, and a report on its condition obtained, as prescribed in the Regulations for Electrical Installations issued by The Institution of Electrical Engineers.

4.3 Protection of the installation

Overcurrent protection

The methods employed for automatic disconnection of the electric supply in the event of an 'earth leakage' were discussed in the previous chapter. This section considers the means of protection we can use against 'overcurrent'.

Most people understand very well what it means to 'overload' a circuit. If you have a perfectly sound ring final sub-circuit and you plug in two 3 kW electric fires the circuit will behave normally, plug in four and the protective device will operate and the circuit will be disconnected from the supply. This is because an overcurrent exceeding the rated value of the protective device is flowing in the circuit. For electrical conductors the rated value is the current-carrying capacity (see Part Two of the IEE Regulations), and if this is exceeded the temperature of the cable will rise and if allowed to continue would result in damage to the insulation, joints, terminations, or surroundings of the conductors (see Regulation 433–1).

A second way that overcurrent can flow in a circuit is when a 'short circuit' occurs. If the phase and neutral conductors in a circuit were inadvertently shorted out, then because of the negligible impedance, a current many hundreds of times greater than the design current of the circuit would flow. The protective device must disconnect the supply very quickly to avoid overheating and mechanical stress caused by the magnetic force set up in the conductors.

Control and Protection for the Consumer

Overcurrent protective devices

The protective devices used to detect overcurrent and to disconnect the circuit for protection against it are:

- Semi-enclosed rewirable fuses to BS 3036;
- Cartridge fuses to BS 1361 and 1362;
- High breaking capacity (HBC) fuses to BS 88;
- Miniature circuit breakers (MCB) to BS 3871.

The advantages and disadvantages of the above devices were discussed in Book 1 and the accompanying Information Sheet gave illustrations of a representative selection of these. If a decision has to be made about which of these is to be used in a particular set of circumstances then we must know a little more about them.

Semi-enclosed rewirable fuse This consists of a porcelain or plastic fuse-bridge and base. The fuse-bridge has a pair of brass pins which slot through apertures in the base and push-fit into suitable contacts top and bottom. The pins are connected with a fuse element consisting of tinned copper wire which either passes over a heat resisting pad or passes through a hole in the plastic or porcelain fuse-bridge. When in place the fuse element is semi-enclosed which gives rise to the name. Different sizes of tinned copper wire have different current ratings and the bridges are either colour coded, have the current rating written on them or both. In order that the wrong bridge is not fitted into the wrong base these are made in different sizes. Information Sheet No. 4B gives a selection of the more popular sizes of tinned copper for use as fuse elements and a colour code list is included.

Cartridge fuses to BS 1361 and 1362 The BS 1362 cartridge fuse has come into common use in the 13 A plug-top used for the connection of appliances. The fuse element is contained in a porcelain tube fitted with two end caps to which the fuse element is welded or soldered. The cartridge fuses too are colour coded, the 13 A one being brown.

Cartridge fuses to BS 1361 can be used as an alternative to rewirable fuses in all the popular makes of switchgear and IEE Regulation 533–4 makes it clear that they are preferable to rewirable types.

High breaking capacity (HBC) fuses Sometimes referred to as high rupturing capacity fuses, these fuses are designed to protect circuits against heavy overloads, and are capable of disconnecting a circuit without damaging the surrounding equipment. HBC fuses are colour coded in exactly the same way as the fuses above.

They consist of a ceramic tube fitted with metal end caps and fixing lugs. The fuse element consists of a silver strip of special design, with a small 'tell-tale' piece in the centre. This has a low melting point and indicates when the fuse has blown. The element is surrounded by purified silica which has the dual role of keeping the element cool under load conditions and preventing an 'arc' occurring under fault conditions.

> **Information Sheet No. 4B The size of tinned copper wire used with semi-enclosed fuses.**
>
> | Current rating (A) | Diameter of wire (mm) | Colour code |
> |---|---|---|
> | 5 | 0.20 | White |
> | 15 | 0.50 | Blue |
> | 20 | 0.60 | Yellow |
> | 30 | 0.85 | Red |
> | 45 | 1.25 | Green |
> | 60 | 1.50 | Purple |

Miniature circuit breaker (MCB) These are mechanical devices for making, carrying and breaking currents under 'normal' conditions. They must also be capable of making, carrying for a short time and breaking currents under specified 'abnormal' conditions, for example short circuit conditions.

The MCB is normally switch operated, and is turned on or off manually. Under overload or fault conditions it will break the circuit automatically and will require resetting once the circuit has been made good. Two methods of detecting overcurrent are used and very often both are incorporated into the one device.

(1) Thermal – this method relies on the overcurrent heating a bimetal strip. If the overcurrent, due to an overload of the circuit, persists, then the bimetal strip will bend thus disconnecting the circuit. There is a specific time lag with this type of protection which allows it to discriminate between sustained and transient overloads.
(2) Magnetic – the overcurrent release is operated by the magnetic effect of the line current flowing in a coil consisting of a few turns of heavy gauge copper. The magnetic effect attracts an iron section which trips the contacts immediately a heavy overcurrent flows, and is therefore ideal for protection against overcurrent due to short-circuit conditions.

Information Sheet No. 4C shows details of the construction of the above methods of protection against overcurrent.

Fusing factor

Very often fuses for overload protection are replaced by people other than those skilled in electrical installation work. In a situation such as this, Regulation 533–2 says that the fuse should be marked or have adjacent to it an indication of the type of fuse to be fitted. The regulation goes on to say that the fuse link should be of such a type that there should be no possibility of inadvertent replacement by a link having the intended nominal current but a higher 'fusing factor' than that intended.

We are aware that different types of protective devices have their own individual characteristics and Figures 8 to 15 in Appendix 8 of the Regulations show time/current characteristics for some of these. Even fuses with the same current rating do not behave in the same way if their fusing factor is different.

The fusing factor of a fuse is a way of knowing how it will perform and can be worked out by the following formula:

$$\text{Fusing factor} = \frac{\text{fusing current}}{\text{current rating}}$$

Where the 'fusing current' is the minimum current causing the fuse to blow, and the 'current rating' is the current that the fuse will carry continuously without blowing. For example the time/current characteristics for a 30 A semi-enclosed rewirable fuse show that 58 A must flow before disconnection of the circuit takes place. Therefore the fusing factor of that particular fuse is 58 divided by 30 or 1.93.

Discrimination

Overload protection is related to the current-carrying capacity of the conductors in the circuit. This current-carrying capacity can be reduced by changing to a different type of cable, reducing the csa of the cable, an increase in ambient temperature or altering the method of its installation where its ability to lose heat is affected.

Where any reduction in the current-carrying capacity of conductors takes place, the Regulations require overcurrent protection to be installed (see Regulations 473–1 and 473–5). Figure 4.4 shows an example of how the protection might be arranged.

It can be seen from the example given that in the normal installation several protective devices are connected in series and it is essential that they operate in the correct order to avoid disconnection of circuits having no faults.

If the two fuses marked X and Y are in series as in Fig. 4.4, then

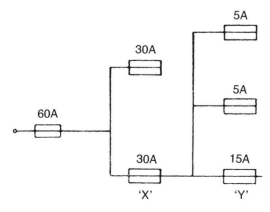

Fig. 4.4 Arrangement of overcurrent protection

Information Sheet No. 4C Protective devices.

1. High breaking capacity (HBC) fuse

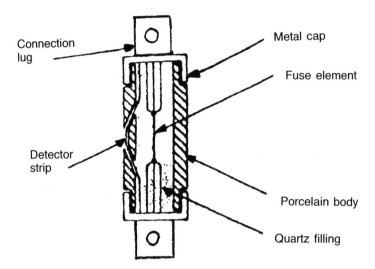

2. Miniature circuit breaker (MCB): thermal and magnetic

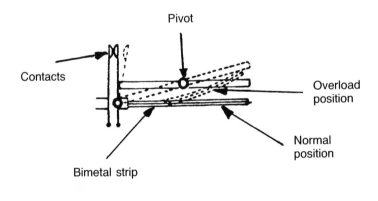

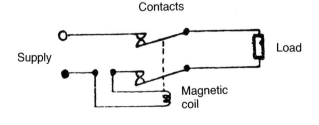

Information Sheet No. 4C (cont.)

3. Semi-enclosed rewirable fuse

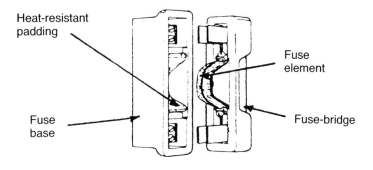

4. Cartridge fuse to BS 1361 and 1362

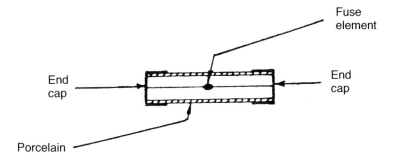

'discrimination' would depend on whether the fault current would be completely interrupted by the smaller fuse Y, before the total energy in the larger fuse X melted the fuse element.

Because of the individual characteristics of the different protective devices, great care should be taken when selecting them for use in an installation. By mixing different types of protective devices it would be possible to create a situation where the protective device with the larger current rating operated before the device with the smaller current rating.

Test 4

Choose which of the four answers is the correct one.

(1) A switchfuse combines the functions of:

(a) switching and isolating; ✓
(b) isolation and protection;
(c) disconnection and isolation;
(d) protection and fusion.

(2) 'Overcurrent' can flow in circuit for two main reasons, these are:

(a) dampness and corrosion;
(b) hysteresis loss and eddy currents;
(c) overload and short circuits; ✓
(d) lagging or leading currents.

(3) The initials MCB stand for:

(a) main circuit breaker;
(b) main current breaker;
(c) miniature current breaker;
(d) miniature circuit breaker. ✓

(4) The fusing factor of a fuse is a way of knowing how it will perform and can be found by dividing the:

(a) fusing current by the current rating of the fuse; ✓
(b) current rating of the fuse by the fusing current;
(c) current rating of the fuse by the design current;
(d) design current by the current rating of the fuse.

(5) Where any reduction in the current-carrying capacity of conductors takes place the IEE Wiring Regulations require:

(a) the equipotential bonding zone to be extended;
(b) that the cpc is twice the size of the phase conductor;
(c) overcurrent protection to be installed; ✓
(d) short circuit protection to be installed.

Chapter 5
The Installation of Cable Tray and Ladder Rack

5.1 Cable tray

Reasons for choice

If it is required to run several sheathed cables such as MIMS or PVC/SWA/PVC along a common route then the time spent clipping and saddling the individual cables can be saved by the installation of a cable tray. If careful thought is given to the positioning of the cable tray, then the pipe work, ducting and equipment of other services can be avoided, thus facilitating the installation of the electrical services.

The cable tray consists of a perforated metal channel which once installed can have cables fastened to it by the means of cleats or cable ties (see Book 1). It is manufactured in widths of 50 mm to 900 mm and in a variety of different types and finishes.

5.2 Determining the size of cable tray

Allowing for grouping factor

Care must be taken when working out the size of the cable tray to be used, as the bunching or grouping of cables affects their current-carrying capacity. Table 9b of the IEE Regulations gives correction factors to be applied to the tabulated current-carrying capacity of the cables if grouped. If, however, the cables are fixed in such a way that no less than twice the diameter of the cables is left between them, then no grouping factor need be applied (see Fig. 5.1). Different types of cable tray are suitable for different jobs so it might be a good idea to look at some of these.

5.3 Types of cable tray

Standard cable tray

Made from perforated sheet steel the standard cable tray consists of a simple flat tray with a turned up edge. It is available in widths varying from 50 mm to 900 mm and is most suitable for the installation of lightweight cables such as MIMS cables or the smaller sizes of PVC/SWA/PVC. The method of connecting lengths of tray together is by means of a reduced end section

Cable Tray and Ladder Rack 53

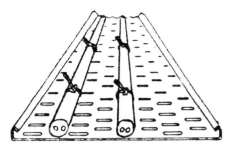
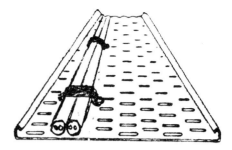

Spaced (space between is twice the diameter of the cable)

Grouped or bunched

Fig. 5.1 Grouping of cables

which fits into the the end of a standard width of tray (see Information Sheet No. 5A). It comes with a full range of accessories such as Tee's, four-way intersections, reducers bends etc. (see Information Sheet 5B).

Heavy duty cable tray

Like the standard cable tray this too is manufactured from perforated sheet steel; however, this is of a heavier gauge and the flanged edge is deeper. Heavy duty trays are made to the CEGB specification 12171 and are suitable for medium duty installation work. There is a full range of accessories for this type of tray and it comes in widths from 150 mm to 600 mm.

Returned flange cable tray

There are a number of different patterns of this type of cable tray varying from a simple returned flange to the heavy duty types. The returned flange gives the tray additional strength and therefore it can span greater distances without support than the standard cable tray. Most of the return flange types of tray are joined together by couplers which fit over the outside of the tray flanges and are quick and easy to use (see Information Sheet No. 5A).

Cable tray finishes

The cable tray can be obtained in a number of different finishes:
- Unfinished sheet steel;
- Red oxide undercoat;
- Yellow chromate undercoat;
- Hot dipped galvanised;
- Epoxy resin coated;
- Plastic coated.

Information Sheet No. 5A Cable tray.

1. Standard cable tray

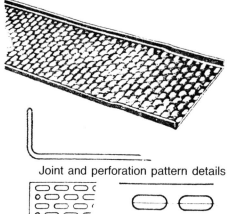

Joint and perforation pattern details

Perforation pattern for 51 mm tray.

2. Returned flange cable tray

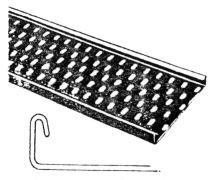

Metal return flange coupling and perforation pattern details

Cable Tray and Ladder Rack

Information Sheet No. 5B Standard accessories for cable trays.

1. Unequal tee

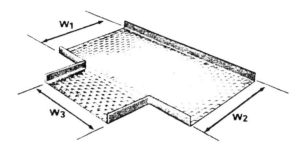

2. Four-way cross piece

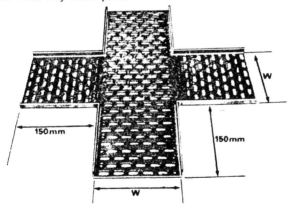

3. Straight reducer

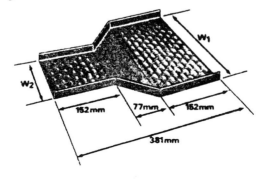

The hot dipped galvanised finish is without doubt the most popular of the above. It is suitable for use in high or low ambient temperatures and the finish can withstand hard knocks without detriment. Unfinished, red oxide and yellow chromate are not suitable finishes unless given a final coat of paint; the yellow chromate has the added advantage of being fire retardant. A number of different plastic coated finishes are available; nylon which is hard and withstands knocks is good where protection against inorganic salts, alkalis, organic solvents and organic acids is required. It is not recommended, however, for use where it might come into contact with chlorinated solvents or inorganic acids. Polythene and PVC finishes are resistant against acids and alkalis, but have poor resistance to organic solvents. Neither polythene nor PVC finishes stand up to hard knocks. Epoxy resin finished cable tray is more expensive than other finishes; however, it has a hard coating that can withstand knocks and is resistant to acids, alkalis, ammonia and salt air. The finish is almost non-flammable and can operate in temperatures between $-50°C$ and $70°C$.

5.4 Installation of cable tray

Fabrication

If full use is made of the wide range of accessories available for the cable tray then little trouble will be experienced with its installation. If certain pieces have to be fabricated on site, then provided simple rules and procedures are followed these can be accomplished with little effort. Information sheet No. 5C shows a cable tray bending machine being used to form a 90° inside riser (a bend with the return edge on the inside).

It will be seen that with the use of a bender the task is very simple indeed; however, 90° risers can be fabricated with the use of simple hand tools and Fig. 5.2 shows how this can be done with the use of a short piece of steel bar with a suitable notch in one end. The notch is placed over the edge of the tray and the bar is used to 'crimp' each side in turn until the required angle is achieved. Simple joints, Tee's, and reducers can be fabricated on site; however, the use of manufactured accessories is recommended in order to save time spent on the job.

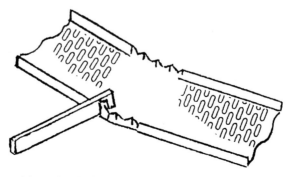

Fig. 5.2 Bend formed by 'crimping'

Cable Tray and Ladder Rack 57

Information Sheet No. 5C Forming a right-angle bend in cable tray.

1. Place tray in bender, locate the edge of the tray into grooved wheel formers

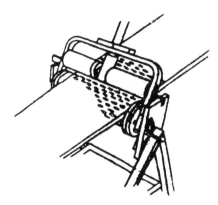

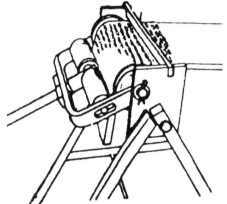

2. Pull the arm of the bender down until desired angle is reached

3. Bend should be formed as shown

Information Sheet No. 5D Ladder racking.

1. Standard length of ladder racking

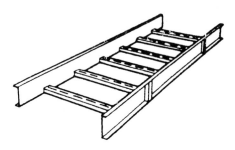

2. Curved section

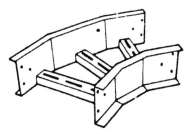

3. T-junction

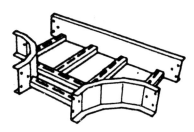

4. Four-way section

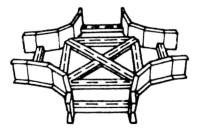

Installation

It is essential that the cable tray is 'spaced off' the surface on which it is to fixed. There are two main reasons for this:

- space is required to push the bolt through if cable cleats are to be used and if cable ties are being utilised then these will need to be slotted in and round the cable;
- it is a good idea to have an air space all round the cables as this reduces the build up of heat.

While standard cable tray can be installed using a simple 'top hat section' bracket, heavy duty cable trays will require something more substantial. Manufacturers such as BICC, Unistrut and Cablock produce special channel together with a whole range of accessories which make installation so much easier and ensure that the tray is well supported and secure.

5.5 Ladder racking

Types of ladder racking available

In situations where there are a number of large cables to be installed along a common route, then serious consideration should be given to the installation of cable ladders. Of robust construction, this form of cable support is ideal in situations such as switch rooms and transformer chambers where large numbers of heavy duty cables come together.

There are a number of different types available, the one shown on Information Sheet No. 5D is constructed of 2 mm gauge mild steel and has a large number of accessories available. The rungs are slotted to take cable cleats and erection is a simple mechanical assembly job.

Test 5

Choose which of the four answers is the correct one.

(1) The bunching or grouping cables on cable tray affects the:

(a) size of cable retainer;
(b) current carrying capacity of the cables;
(c) the radius of the bends;
(d) ambient temperature of the surroundings.

(2) When a returned flange is used on cable tray it:

(a) enables cables to rise up the wall;
(b) lets cables enter the distribution board;
(c) gives the tray additional strength;
(d) makes joining the tray easier.

(3) If cable tray is to be used in damp situations the finish should be:

(a) unfinished sheet steel;
(b) red oxide;
(c) yellow chromate;
(d) hot dipped galvanised.

(4) Ladder rack or cable ladders should be used for:

(a) reaching high level cables;
(b) supporting lightweight cables;
(c) storing access equipment on;
(d) supporting heavy cables.

(5) To avoid having to use the grouping factor when calculating the required size of cables these should be spaced on tray at:

(a) not less than two diameters of the cable between them;
(b) a 20 mm space should be left between them;
(c) the width of a cable cleat;
(d) not less than one diameter of the cable.

Chapter 6
The Installation of Cable Trunking and Under Floor Ducts

6.1 Metallic trunking

Reasons for choice

Trunking is used to accommodate a large number of small cables, or to install cables which are too large to be drawn into conduit. IEE Regulation 529−7 states that the number of cables drawn into, or laid in an enclosure of a wiring system shall be such that no damage is caused either to the cables or to the enclosure. It will be noticed that Table 12A of the Regulations does not indicate a cable size larger than 10 mm^2 for installation in conduit systems and in Table 12B conduits greater than 32 mm in diameter are not listed. Although conduits larger than this are available, in practice it is cheaper and more convenient to install cable trunking if a capacity greater than 32 mm is required.

Cable trunking offers a highly versatile and adaptable system of cable installation. It provides good mechanical protection to cables, so it is entirely suitable for installations in workshops or industrial premises. The standard trunking with its removable lid means that circuits can be added or removed with relative ease and, provided the regulations on segregation of different types of circuit are complied with, the cables need only be of the single PVC insulated type. This type of enclosure can be used as a protective conductor and the manufacturers maintain that the connectors provided are suitable to achieve this. However, for installations where high fault currents could occur then additional copper links fitted across the joints or the provision of a separate CPC is recommended. Under no circumstances, however, should metallic trunking be used as part of a protective earth and neutral (PEN) system (for definition of PEN systems see Book 1).

Disadvantages of the metallic cable trunking are that it is quite expensive compared to some other wiring systems, and it requires a certain amount of skill in its installation and is therefore labour intensive. Also, it can be subject to corrosion under certain conditions, and it is difficult to make it gas- and waterproof in the same way that conduit installations can be.

6.2 Determining the size

Trunking is available in various sizes from 38 mm by 38 mm section to 300 mm by 150 mm section, and is generally supplied in 2.5 m or 3 m lengths together with one connector complete with nuts and bolts. Standard bends, tees, flanges and other accessories are available (see Information Sheet No. 6A).

The size of trunking required can be worked out from the Tables provided for that purpose in the IEE Wiring Regulations Appendix 12. For each of the cables that are going to be installed a factor for that particular size of cable is given in Table 12E. The factors are added together and compared to the factors for trunking given in Table 12F. The size of trunking that is most suitable for use with these cables is the one whose factor is equal to or greater than the sum of the cable factors. For example if we had to install ten 4 mm^2 cables, ten 6 mm^2 cables, and ten 10 mm^2 cables, the total factor for the cables using the stranded type of cable would be 152 + 229 + 363, giving a total factor of 744. A look at Table 12F would show that a 75 × 25 mm trunking at a factor of 738 would be just too small and that the 50 × 37.5 mm trunking at a factor of 767 would in fact be more than adequate so this size would be used. For sizes of cables and trunking not given in the above tables the number of cables installed should be such that the resulting spacing factor does not exceed 45%. The space factor in this case is the ratio of the sum of the overall cross-sectional area (csa) of the cables (including cable and sheath) to the internal csa of the trunking. This is calculated as follows:

$$\frac{\text{Sum of the CSA of the cables}}{\text{Internal CSA of the trunking}} \times 100\%$$

6.3 Installation of metallic cable trunking

Cable trunking is first and foremost a metal enclosure for the protection from mechanical damage of cables to be installed in it, therefore it should be installed in such a way as to afford continuous protection for the cables and allow safe and easy installation or withdrawal of such cables. The trunking system should be fully fabricated and erected before any attempt is made to install the cables into it (Information Sheet No. 6B shows the fabrication of a right-angled bend). Secondly, because the metallic trunking can be used as a protective conductor it is essential that the joints formed by connectors, flanges, bends, reducers and other fittings are carried out in such a way as to ensure mechanical and electrical continuity throughout its length. It is these two main points that concern us when considering the installation of metallic trunking, so it may be useful to look at some of the requirements for these to ensure a safe and sound installation.

- Trunking ends must be cut square to ensure a good fit into accessories.
- Any burrs must be removed with a flat file, to avoid abrasion of cables.
- Where cables are likely to come into contact with trunking edges, grommet strip should be fitted.
- All screws on connectors and fittings must be securely tightened.
- Round-headed fixing screws only should be used to avoid snagging the cables when installed.
- Any bends should have a diameter such that the radius of the cables contained within them complies with Regulation 529–3.

Cable Trunking and Under Floor Ducts 63

Information Sheet No. 6A Standard trunking accessories.

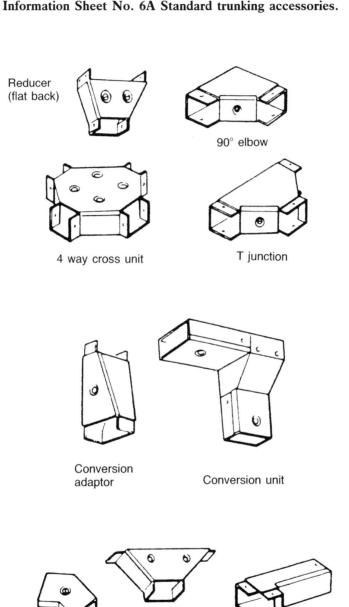

Reducer (flat back)

90° elbow

4 way cross unit

T junction

Conversion adaptor

Conversion unit

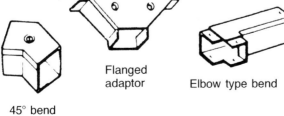

45° bend

Flanged adaptor

Elbow type bend

Information Sheet No. 6B Fabrication of right-angled bend.

(a) Mark out 90° angle at position of bend on both sides as shown.

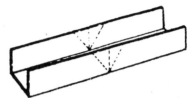

(b) Cut away unwanted metal on both sides.

(c) File all edges smooth and bend into shape.

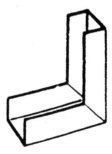

(d) Make a fishplate out of scrap metal and mark and drill holes. Mark out trunking from fishplate as shown. Drill trunking, file off any burrs and assemble with nuts and bolts.

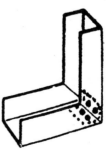

Cable Trunking and Under Floor Ducts

- Recommendations regarding fire barriers where trunking passes through walls and floors should be observed.
- Barriers should be fitted at a minimum of every 5 m on vertical trunking to avoid heat collecting at the top.
- Holes cut through the fabric of the building to accommodate trunking, should be made good with incombustible material to the full thickness of the hole.
- Holes for the connection of conduit accessories to the trunking must be cut or punched using appropriate tools.
- Any conduits shall be secured by means of a bush and connector and any unused conduit entries should be blanked off.
- Cables in vertical trunking exceeding 5 m in length should be supported with intermediate cable supports.
- Cables installed in trunking should be held in place with suitable cable retainers.
- The correct space factor for that size of trunking must not be exceeded.
- All lids for the trunking and its fittings and accessories should be fitted and securely fastened in place.
- In accordance with Regulations 523–7 to 523–15, care should be taken where there is a likelihood of water or moisture.

6.4 Types of metallic cable trunking

Standard trunking

Standard trunking is usually made of zinc coated low carbon sheet steel, although aluminium types are available for certain applications. Standard trunking comes in various finishes as follows:

- Painted or sprayed grey enamel;
- Painted or sprayed silver enamel;
- Hot dipped galvanised coating.

The hot dipped galvanised finish is thicker than the electrically deposited zinc coating and is very durable.

All the standard fittings and accessories are available, such as elbows, tees, reducers, flanges and adapters, and it is only in exceptional circumstances that the electrician has recourse to prefabricate these (see Information Sheet 6B).

Lighting trunking

Without doubt the biggest increase in recent times for the use of cable trunking is the widespread use in industrial and commercial premises of the lighting trunking. Easy to install with the use of specially designed hangers, it can span the large distances between roof supports of the modern prefabricated premises. Not only does it give mechanical protection for the cables but it provides a means of mounting the luminaires in neat straight

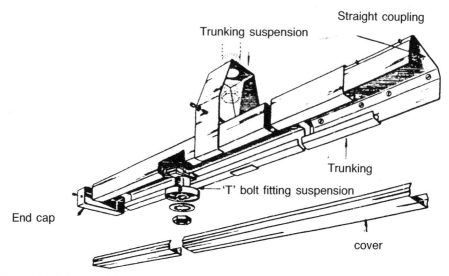

Fig. 6.1 Lighting trunking

rows, and reduces the number of fixings required to the fabric of the building. The trunking is manufactured from galvanised sheet steel and has a folded return along the bottom edge giving it greater strength and providing a method of suspension for the luminaires and the trunking hangers. The most popular size is 50 mm × 50 mm and comes in 4 m lengths. An example of lighting trunking is shown in Fig. 6.1 together with suspension bracket, straight coupling, end cap, fitting suspension, and snap-in cover.

Skirting trunking

As its name implies, skirting trunking is fixed in place of the normal skirting board. Its main use is confined to the outer perimeter of rooms where there is a call for a large number of outlets for small power, telephone and computer outlets. The metal version is very expensive and its use really needs to be justified before choosing this as an installation method. Information Sheet No. 6C gives an example of this type of trunking and shows how it is connected to under floor trunking.

Bench and dado trunking

Where there is a need for multiple electrical service outlet points at bench height then this form of trunking can be considered. It is ideal for use in laboratories or data processing rooms where outlets for various voltages, telephones and computer networks are required. Some types can be continued on round rooms in the form of a dado providing services to desks, computer servers, workstations etc.

It comes in multi-compartment types to provide segregation of the dif-

Information Sheet No. 6C Skirting trunking.

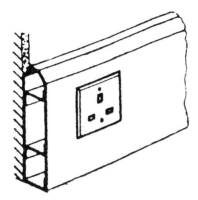

1. Skirting trunking with lid fitted and 13A socket outlet in place.

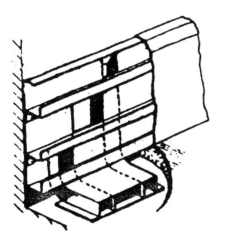

2. Showing lid removed and how under-floor trunking interconnects with it. Notice the three compartments for the different categories of wiring.

3. Pitch fibre ducting showing how the adaptor screws into the crown of the ducting.

ferent services, and can be obtained in a number of attractive finishes and styles.

6.5 Under floor trunking and cable ducts

In some large buildings under construction it is sometimes found practical and economical to provide a network of cable ducts or trunking in the concrete floor. One advantage of this is that in large commercial buildings there are often changes of tenancy of individual office suites which may entail alterations of the layout of the areas. This can be effected more easily if there is a system of ducts or trunking, particularly in the large open plan offices favoured today. There are a number of different types available.

Flush floor trunking

This type of trunking is seated in a bed of concrete and is levelled off at 'finished floor level'. The lid of the trunking takes the form of a shallow tray into which tiles, wood blocks or even chequer plate can be fitted to suit the proposed floor finish. Access is possible throughout its length and it is much used in hospitals and buildings of this nature. Made from zinc coated sheet steel it comes in 2 m lengths and is available in a number of different sizes and can have a number of separate channels for different services (see Information Sheet No. 6D).

Under floor trunking

Trunking of this type is installed directly onto the structural slab of new buildings, and the screed is poured over it until it is completely submerged. Access is obtained only at junction boxes or where the trunking rises up the wall to go to a distribution board or join up with skirting trunking. Made from heavy duty galvanised sheet steel it comes in 2.5 m lengths. It can be of either single or multi-channel section and access to outlets is by connecting sleeves mounted on brass plates (see Information Sheet No. 6D).

Flush duct trunking

Like the flush floor trunking this type of trunking is set in concrete so that its lid is level with the finished floor. The floor finish is then placed over it. Unlike the flush floor trunking, access can be made at any point simply by drilling down through the floor finish and the trunking lid, and screwing in a conduit connector which has a special spigot end which fits into the lid (see Information Sheet No. 6D).

Information Sheet No. 6D Under floor trunking and ducts.

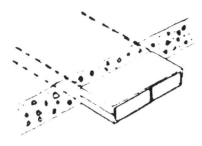

1. Under floor trunking

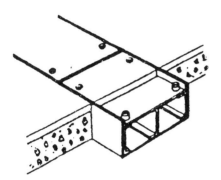

2. Flush floor trunking

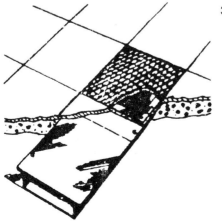

3. Flush duct trunking

Pitch fibre ducting

One such system comprises semicircular pitch-impregnated fibre ducting which is laid on the floor before finishing. The system is available for various depths of floor screed and access is obtained by drilling holes in the floor at required positions on the route. Cables installed in cable ducts must be of the sheathed or armoured types, and CPCs should be provided. Information Sheet No. 6C shows a cross-section of this type of trunking and shows an adapter going through the floor screed into the trunking for tapping off at the required position.

Multibeam trunking

A number of buildings such as warehouses and supermarkets are being constructed with roofs comprising of sheet material rather than the traditional tiles or slates. The purlins forming part of this roof structure can, if it suits the architect's purpose, also form part of the electrical installation. 'Multibeam' purlins (see Fig. 6.2) can accommodate cables for lighting and small power in a recess formed by the way the steel is folded to give strength to the beam; a PVC cover much like that used with lighting trunking completes the installation. The metal purlins will be bonded as part of the equipotential bonding system of the installation and permission must be obtained from the architect before any holes are cut in the purlin as this could reduce its weight-carrying properties.

Suspended floor installations

Many computer suites and data processing rooms have suspended floors in order to accommodate more easily the proliferation of services into the area. A number of manufacturers have seen the need for a satisfactory method of terminating cables in this sort of situation and Fig. 6.3 shows examples of how this is achieved.

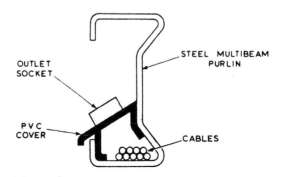

Fig. 6.2 'Multibeam' trunking

Cable Trunking and Under Floor Ducts 71

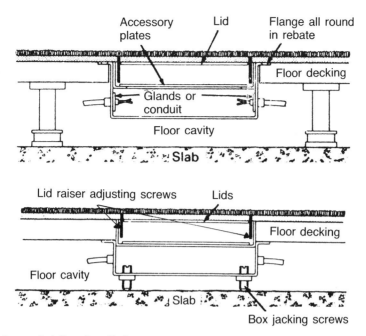

Fig. 6.3 Suspended floor installations

6.6 Bus-bar trunking systems

Trunking is often used to enclose bus-bars. These bus-bar systems may be roughly classified into two groups.

Overhead bus-bar trunking

The metal clad overhead bus-bar system is often used for three phase distribution in factories to feed a number of machines. The usual arrangement consists of zinc coated sheet steel trunking finished in grey stove enamel, containing copper or aluminium bus-bars mounted on insulators. At intervals, for instance every metre, tapping off points are provided to which a fused unit can be fitted. The fused units consist of some means of making contact with the bus-bars, usually some type of socket or clamping arrangement. Connection from the fused unit to the equipment is made either by flexible connections, cable in conduit, mineral insulated cables etc.

The initial cost of the overhead bus-bar trunking is high; however, once installed it provides a highly flexible system to which additions and alterations can be carried out quickly and easily. Information Sheet No. 6E shows an example of this trunking with a typical plug in type fused unit.

Information Sheet No. 6E Bus-bar trunking.

1. Overhead bus-bar trunking

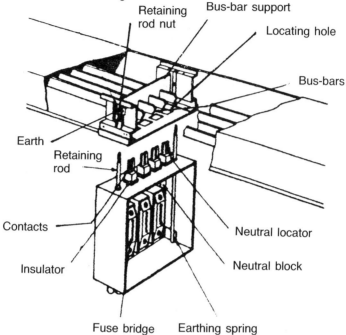

2. Rising-main bus-bar trunking

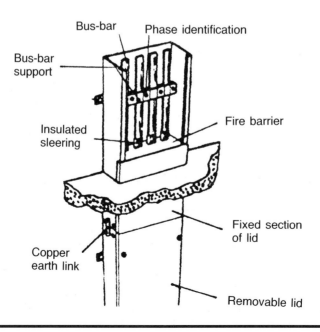

Rising main bus-bar trunking

For electrical installations in large multi-floor buildings, bus-bar trunking is sometimes used for vertical rising mains. It consists of a zinc coated sheet steel case finished in grey stove enamel. The sections are joined by the use of connectors complete with plated steel screws, copper earthing links and shake proof washers in much the same way as standard cable trunking. The trunking contains either copper or aluminium bus-bars, often extruded in PVC insulation and colour coded to help identification of the phases. These are mounted on insulators made of laminated insulating material. The sections of bus-bar are connected by solid copper or aluminium links, but in extremely long runs, joints consisting of flexible braided tape are included at certain points to take up any variations in length due to temperature change. Thrust blocks are positioned at the base of each set of bus-bars to take any downward pressure. The rising bus-bars are usually fed at the bottom of the trunking run, and provision will be made by the manufacturers for the entry of PILC, PVC/SWA/PVC, or MIMS cables. Tapping off points are provided at each floor level to supply local distribution boards and these can be be bolted directly to the trunking if required. It is important that fire barriers are placed at each point that the trunking passes from one floor level to another (see Information Sheet No. 6E) and that the hole which has been cut for it to pass through is made good with non-combustible material to the full thickness of the hole. The Regulations call for the fitting of barriers every 5 m to prevent the trunking acting like a flue and the hot air rising and collecting at the top of the trunking and possibly causing damage to the insulation of the cables.

6.7 Plastic trunking

Many of the trunking types discussed above can be obtained in High Impact PVC to BS 631. These are suitable for many different applications in domestic commercial or industrial situations and have the added advantage of being light in weight, easy to cut and prefabricate. The IEE regulation 521–12 asks that they comply with British Standards fire tests on building materials and structures BS 476 Part 5 and in addition they are acid and corrosion resistant and can be obtained in more attractive colours than the metallic types.

This type of trunking in certain circumstances suffers from a number of disadvantages. It expands and contracts with changes in temperature, and it cannot be used as a CPC so an additional cable must be installed for that purpose. Although it is quite tough it is not as robust as steel trunking. Because of its lack of strength more fixings will be required per length than steel trunking and it cannot span the same sort of distances that steel trunking is capable of.

Having mentioned the above disadvantages it has got to be said that in the right circumstances and under suitable conditions, the speed of erection of this type of trunking can sometimes outweigh some of the disadvantages and it should not be dismissed as a valid form of wiring system.

6.8 Multi-compartment trunking

It is a requirement of the IEE Wiring Regulations that cables of certain different 'categories' shall not be contained in the same conduit, trunking, ducting or multi-cored cable. As a result of this most of the types of trunking and ducting mentioned above can be obtained with more than one compartment (see Information Sheet No. 6C).

6.9 Segregation of circuits

Circuits are split into three categories:

Category 1 circuit A circuit operating at low voltage and supplied directly from the mains supply (other than fire alarm or emergency lighting circuits).

Category 2 circuit A circuit for telephones, radio or television, sound distribution, intruder alarms, bell and call and data transmission circuits, which are supplied directly from a safety source complying with Regulation 411–3 (other than fire alarm or emergency lighting circuits).

Category 3 circuit A fire alarm circuit or an emergency lighting circuit.

Regulations 525–1 to 525–9 lay down the requirements for segregation and these are briefly:

- Cables of Category 1 circuits must be segregated from Category 2 circuits unless the latter are insulated in accordance with the regulations to the highest voltage present in the Category 1 circuits.
- Under no circumstances must Category 1 circuits be installed in the same conduit, duct, trunking, multi-core cable, flexible cable or flexible cord as Category 3 circuits.
- Multi-compartment trunking can be used to house the different categories of circuit provided the partition used to separate the different categories is strictly in accordance with the above regulations.
- Where the Category 3 circuits are wired in mineral insulated cable and installed in a common duct or trunking, then the above partitions are not normally required; however, if this is the case, then these should be rated at the 'exposed to touch' conditions as given in Tables 9J1 and 9J3 of Appendix 9 of the IEE Wiring Regulations.
- Category 1 circuits and Category 2 circuits can be in the same multi-core cable provided the latter are insulated to the highest voltage present in the Category 1 circuits or are separated from the Category 1 circuits by an earthed metal braid of the equivalent current-carrying capacity to that of the cables of the Category 1 circuits. Where these two categories share a common terminal or connection box the cables and connections shall be separated by rigidly fixed screens or barriers (Regulation 525–7) unless they are mounted on separate and distinct terminal blocks clearly marked to indicate their function (Regulation 525–8).

6.10 Electromagnetic effects

Very often circuits installed in a metallic trunking system will be required to leave the trunking either to continue in a conduit system, enter into a terminal box or simply to gain access into switchgear. Care should be taken under these circumstances to ensure that all conductors of an alternating current circuit are enclosed in the same steel conduit or enclosure when entering or leaving the trunking system. Failure to carry out this simple precaution can result in excessive heat being generated in the surrounding metal work.

When a conductor carries an alternating current it sets up an alternating magnetic field around itself. This magnetic field is much stronger when it is surrounded by ferrous metal such as conduit connectors or steel plates, and eddy currents induced in the metal work can cause it to heat up. This overheating can be avoided by taking the above precautions and ensuring that the phase and neutral conductors of a single phase circuit are in the same enclosure. The reasons for this are that the currents in the phase and neutral conductors will be equal and will provide equal and opposite ampere-turns so that the magnetic flux will be negligible. In the same way the four conductors of a three phase and neutral supply should be kept together to ensure that the magnetic flux is kept to a minimum. The situation can be further improved by slotting the metal boxes or steel plates where cables (and in particular single core cables) enter, thus breaking the magnetic path and reducing the eddy currents.

Test 6

Choose which of the four answers is the correct one.

(1) Holes cut through walls to accomodate trunking shall be made good to the full thickness of the wall with:

(a) combustible material;
(b) pieces of polystyrene;
(c) incombustible material;
(d) old newspapers.

(2) Where vertical rising trunking passes through walls and floors it should be fitted with internal barriers to prevent:

(a) the loss of heat;
(b) cold draughts;
(c) the spread of fire;
(d) unauthorised access.

(3) Standard metallic trunking is usually made from:

(a) low carbon sheet steel;
(b) high tensile sheet steel;
(c) wrought iron;
(d) galvanised iron.

(4) Bus-bar trunking systems may be roughly classified into two main groups, these are:

(a) vertical and horizontal;
(b) overhead and rising main;
(c) perpendicular and horizontal;
(d) under floor and flush duct.

(5) Category 3 circuits must not be placed with Category 1 and 2 circuits under any circumstances, unless wired in:

(a) double insulated cable;
(b) butyl insulated cable;
(c) heat resistant PVC cable;
(d) mineral-insulated cable.

Chapter 7
The Heating of Water by Electricity

7.1 Types of water heater

An electric water heater of the resistance type generally consists of an electric element encased in a metal sheath which is in contact with the water to be heated. Apart from kettles, they can be roughly classified into five groups:

- non-pressure or free outlet types;
- pressure type (local storage);
- pressure type (central storage);
- immersion heaters;
- instantaneous water heaters.

Non-pressure type

This type has a cylindrical container with inlet and outlet pipes, the inlet being connected to the main water via a valve and the outlet left open. The heating element and the thermostat are in the bottom of the container. When the inlet valve is opened, the heated water is displaced by an equal volume of cold water and forced out through the outlet pipe. This open or 'free outlet' spout acts as a vent and relieves pressure which would otherwise be set up by the expansion of the water on heating. It must, therefore, never be closed or joined to a tap. Baffles are fitted to prevent incoming cold water mixing with the heated water and an anti-drip device prevents the hot water, which expands at the rate of 2% per 100°C, from constantly dripping from the outlet spout. This particular pattern of non-pressure water heater, known as the free-outlet type, is used to supply hot water to one position only. In this case, it is mounted near to the sink or wash-hand basin where the hot water is required.

The heating element and thermostat are fitted into tinned copper tubes which are fixed on an element plate to facilitate renewal or repair. When installing the heater care must be taken to ensure that there is sufficient room for their easy withdrawal.

Free-outlet heaters suitable for use with wash-hand basins or sinks are as a general rule 6.5 or 13 litres in capacity and those for use with baths 54 or 65 litres capacity. When contemplating installation of the larger types, the weight of the water (1 kg per litre) will have to be considered when choosing the method of fixing (see Information Sheet No. 7A).

Information Sheet No. 7A Water heaters.

1. Non-pressure type

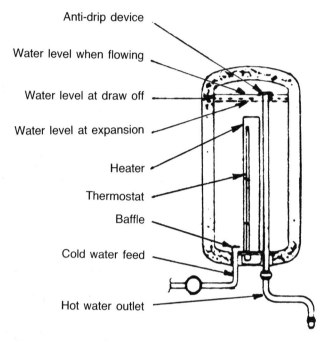

- Anti-drip device
- Water level when flowing
- Water level at draw off
- Water level at expansion
- Heater
- Thermostat
- Baffle
- Cold water feed
- Hot water outlet

2. Pressure type (control storage)

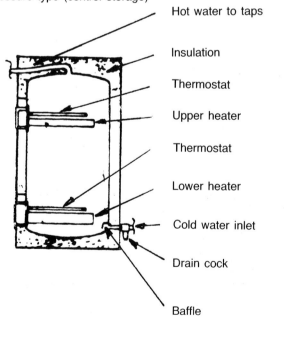

- Hot water to taps
- Insulation
- Thermostat
- Upper heater
- Thermostat
- Lower heater
- Cold water inlet
- Drain cock
- Baffle

The Heating of Water by Electricity

Information Sheet No. 7A (Cont.).

3. Pressure type (local storage)

- Hot water outlet
- Hot water draw-off
- Thermostat
- Heater
- Cold water inlet

4. Cistern type

- Overflow
- Ball valve
- Vent pipe
- Thermostat
- Heater
- Hot water draw-off
- Cold feed
- Drain cock

Pressure type (local storage)

These are mainly used when it is necessary to provide a number of hot water outlets from one water heater. The heater is fed with cold water from a high level cistern connected to the water main and controlled by a ball valve. Outlet points are thus fed under pressure supplied by the head, or vertical height of the cold water available.

The cold water entry to the heater is at its base, and connected to the top of the container of the ordinary type is the hot water outlet/expansion pipe, which terminates in the cistern above the water level – this relieves any pressure set up by the heating of the water. In bungalows where the head of pressure is necessarily low, the hot water outlet can be taken from the bottom.

This type of heater is available in capacities up to 450 litres, with electrical loadings of up to 6 kW. The double skinned walls of the heater contain thermal insulation material to reduce heat loss (see Information Sheet No. 7A).

Pressure type (central storage)

In the 'two-in-one' domestic water boiler, a small thermostatically controlled heater is fitted horizontally at the upper section of the tank and a larger thermostatically controlled heater is fitted at the lower section of the tank. The principal of operation of this system is that the smaller heater is left switched on so that there is a reasonable amount of hot water available at all times. Should large amounts of water be required, say for baths or laundry requirements, then the larger heater is switched on which heats the whole tank. This type of heater is designed to take the place of other types of heater such as solid fuel etc. (see Information Sheet No. 7A).

Cistern type

An alternative water heater, known as the cistern type, has a small built in cold water storage tank controlled by a ball valve. This does not require an expansion pipe, but has an internal vent pipe which discharges into the internal cold water cistern. It can be supplied with cold water direct from the mains (subject to local water authorities agreement); if this cannot be agreed upon it must be fed from a cold water storage cistern in the usual way. Well insulated, these heaters are fitted as high as possible above the level of the taps and can be used very effectively on 'off peak' supplies. They are sometimes installed in blocks of flats, and result in a saving in plumbing costs, due to the elimination of expansion pipes and the use of a common down service for the cold water supply (see Information Sheet No. 7A).

Immersion heaters

Existing water storage cylinders may be fitted with electric immersion heating elements. The system where the hot water cylinder is fed from a cold water cistern placed above it, is really a form of pressure system. The water supplying the cistern is controlled by a ball valve and when the hot water is drawn off the hot water cylinder the cistern replenishes it and the ball valve operates to refill the cistern. In this way a constant pressure or head of water is maintained on the hot water system. To reduce the consumption of electrical energy to a minimum, the tank should be efficiently lagged with a heat insulating blanket. The elements consist of high resistance nickel-chrome heating coil contained in a metal sheath from which it is insulated by impacted mineral oxide. These are fitted with a thermostat for economical control, and are usually rated at 3 kW (see Fig. 7.1).

A dual element type of immersion heater is now available. It consists of a short heater of a rating of about 2 kW and a full length heater of 3 kW all combined in the one unit. The principle behind this is very much the same as the 'two-in-one' heater discussed above, and a built in selection switch allows a choice between a full tank or a partial tank of hot water. Circuit diagrams for the above heaters can be found on Information Sheet No. 7B.

Instantaneous water heaters

Cold water passes through the heater into a very small receptacle containing an electric element which is of a very high rating (usually in the order of 7 to 8 kW). Only a small amount of water is in contact with the element at any one time and so it is heated up almost as soon as it touches it. Unlike in the storage heaters above, the electrical supply is only switched on when the water is flowing through the heater. In the basic models the rating of the

Fig. 7.1 Immersion heater

Information Sheet No. 7B Immersion heaters.

1. Dual element immersion heater

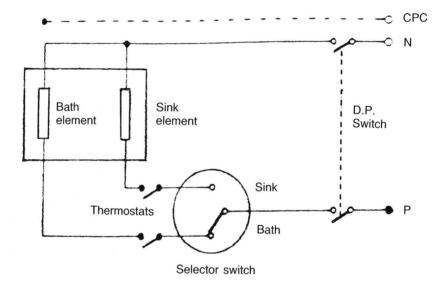

2. Single element immersion heater

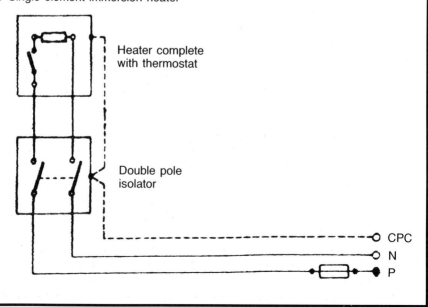

heater is fixed so that the temperature of the water to some extent depends on the rate of flow of the water; in the more expensive types, such things as water and temperature stabilisation, water pressure operated power switches, audio signals and phased shut down facilities are available. In all models there is a built in cut-out which will operate should the flow fall and the temperature become excessive.

Instantaneous water heaters are now available for serving more than one outlet point. Ideal for toilets in service stations cafes etc., they plumb directly into the cold water mains and do not require a header tank. They have 4.7/9.5 kW heat settings and have thermal cut out protection. They have the added advantage that they can be installed below the level of the taps, and the water outlet is via the existing tap/mixer at the sink or wash-hand basin.

7.2 IEE Regulations concerning water heaters

All single phase water heaters or boilers having uninsulated heating elements immersed in water shall have all their metal parts (other than current-carrying parts) connected to the metal water pipe through which the water supply to the heater or boiler is provided, and this pipe shall be effectively earthed independent of the circuits protective conductor (Regulation 554–29).

The water heater or boiler shall be controlled by means of a double pole linked switch which is either incorporated into the apparatus (subject to Regulation 476–8), or is within easy reach of it. If the water heater or boiler is situated in a room containing a fixed bath or shower then any separate switch controlling it must be out of reach of a person using the bath or shower (Regulation 471–39). Under no circumstances should a plug and socket outlet be used to control the heater or boiler (see Regulation 554–30).

Heaters or boilers should have incorporated into them an automatic device to prevent dangerous rises in temperature. However, before the installer connects a heater or boiler of any of the above types, he/she should confirm that all thermally operated devices, single pole switches, non-linked circuit breakers or fuses are NOT fitted in the neutral conductor of any part of the circuit, between the apparatus and the circuit source (see Information Sheet No. 7B for circuit diagrams).

The selection of the heater or boiler should take into account any aggressive conditions of the water or liquid, 'Copper' elements for soft or normal water and 'Aqualoy 825' or its equivalent for extremely aggressive water areas (see the note at the start of Regulation 554–27).

For circuits supplying equipment in rooms containing either a bath or shower all the requirements for bonding etc. referred to in Regulation 471–34 to 39 should be met and this is discussed elsewhere in the book. It should be emphasised, however, that the protective devices and the earthing arrangements for equipment in a room containing a fixed bath or shower shall be such that in the event of an earth fault, disconnection of the supply occurs within 0.4 s.

The type and current-carrying capacity of cables and flexible cords used for the connection of water heaters and boilers shall be suitable for the highest operating temperature likely to occur in normal service (Regulation 523–1). In the case of water heaters and boilers this usually means that the final connection to the heater is made with heat resistant flex such as Butyl or Elastomer Insulated HOFR Sheathed to BS 6500 which is capable of operating in working temperatures up to and including 85°C (correction factors for ambient temperatures above 30°C can be found in the tables in Appendix 9 of the IEE Wiring Regulations).

7.3 Electric heating calculations

An electric current flowing through a conductor generates heat. This property is used to provide heating direct from electrical energy. There are two properties of heat which can be measured: (a) temperature and (b) quantity.

Temperature

Temperature is measured in degrees using a thermometer. Until recently the 'Fahrenheit scale' was used for everyday use and the 'Centigrade scale' used for scientific work. Since the SI unit system has now been adopted, the temperature scale to be used is the 'Celsius scale'. Celsius replaces the name centigrade and one degree centigrade becomes one degree Celsius. The absolute scale of temperature is in Kelvin (K). This is temperature measured from absolute zero, absolute zero being $-273.1°C$. The Kelvin and Celsius intervals are identical but a temperature expressed in °C is equal to the temperature expressed in Kelvins less 273.1. Where it is necessary to convert from Fahrenheit to Centigrade, or centigrade to Fahrenheit then the following formulae are used:

$$°F = (\tfrac{9}{5} \times °C) + 32$$

$$°C = (\tfrac{5}{9} \times °F) - 32$$

Quantity of heat

Heat can be defined as energy in transit between two bodies because of a difference in temperature and is measured in 'joules'.

The joule (J) is the SI unit of energy, which is defined as the work done when the point of application of a force of one newton is displaced through a distance of one metre in the direction of the force (newton metre).

However, the joule is the common form of energy. If a power of one watt is applied for one second the electrical energy expended is one joule (watt second).

If the temperature of one kilogram of water is raised by one kelvin or one degree Celsius, the heat energy expended is 4190 joules.

The Heating of Water by Electricity

Multiples and sub-multiples are used in conjunction with the joule as follows:

10^6 joules = 1 megajoule (MJ)
10^3 joules = 1 kilojoule (kJ)
10^{-3} joules = 1 millijoule (mJ)
10^{-6} joules = 1 microjoule (uJ)

It is useful to remember that:

1 kWh = 1000 × 3600 J
1 kWh = 3.6 megajoules
1 Ws = 1 joule
1 W = 1 joule/second

Specific heat capacity

In defining the unit of heat, water is taken as the standard material. To raise the temperature of different materials by an equivalent amount, less heat would be required.

The actual amount of heat (in joules) necessary to produce a temperature rise of one degree Celsius in one kilogram of a material is called 'the specific heat capacity' of the material.

Table 7.1 gives representative 'specific heat capacities' in J/kg °C of more common materials.

Calculation of heat energy

To calculate the heat energy (Q) in joules we must know the mass of the

Table 7.1 Specific heat capacities

Material	Specific heat capacity (J/kg °C)
Water	4190
Air	1010
Oil (transformer)	2140
Copper	390
Lead	126
Aluminium	946
Cast iron	480

substance (m) in kg, the specific heat capacity of the substance (C) in J/kg °C and the temperature change ($t_2 - t_1$). From this the heat equation formula is derived as:

$$Q = m \times C \times (t_2 - t_1) \text{ joules}$$

Example Calculate the quantity of heat required to raise the temperature of 20 kg of water from 30°C to 100°C

$$\begin{aligned} Q &= m \times C \times (t_2 - t_1) \\ &= 20 \times 4190 \times (100 - 30) \\ &= 5\,866\,000 \text{ J or } 586.60 \text{ kJ} \end{aligned}$$

Efficiency

Although heating by electricity is very efficient, losses do occur especially if the pipes and hot water tanks are not lagged effectively. The efficiency of the system is worked out in the following way:

$$\text{Efficiency} = \frac{\text{Heat required (output)} \times 100}{\text{Heat supplied (input)}}$$

The efficiency is given as a percentage and must be taken into account when carrying out our calculations. For example if we are told that in the example above the efficiency was 85% we would apply it to our calculation as follows:

$$Q = \frac{20 \times 4190 \times 70}{0.85}$$

$$Q = \frac{5\,866\,000}{0.85}$$

$$Q = 6\,901\,176 \text{ J or } 6901.176 \text{ kJ}$$

We can see from this that more heat energy is required to raise the temperature by the same amount.

This illustrates how important it is to lag the water heating system effectively in order to reduce losses and save money.

Test 7

Choose which of the four answers is the correct one.

(1) Water heaters or boilers with current ratings exceeding 3 kW shall be controlled by means of a:

(a) suitably rated simmerstat;
(b) single pole isolator;
(c) double pole linked switch;
(d) socket and plug-top.

(2) The amount of heat required to produce a temperature rise of one degree celcius in one kilogram of a material is called its:

(a) specific heat capacity;
(b) temperature coefficient;
(c) temperature differential;
(d) specific heat loss.

(3) The material of which the heating element of an immersion heater is made is:

(a) tungsten;
(b) copper;
(c) nickel-cadmium;
(d) nickel-chrome.

(4) The two properties of 'heat' which can be measured are:

(a) temperature outside and inside;
(b) temperature and quantity;
(c) initial and final temperatures;
(d) lagged and unlagged cylinders.

(5) If a water heater or boiler is placed in a room containing a fixed bath or shower, then the switch controlling it must be:

(a) within reach of a person using the bath or shower;
(b) of the type with an indicating light;
(c) out of reach of a person using the bath or shower;
(d) manufactured from white plastic.

Chapter 8
Cooking and Space-heating

8.1 The installation of cooker final circuits

Final circuits

Electrical apparatus is connected by cables to the electricity supply and to the associated protective and controlling devices (usually fuses and switches). This arrangement of cables is known as a 'circuit' and circuits which connect current-using apparatus to the consumer unit or distribution board, are called 'final circuits'.

Cooker final circuits (household)

There are a number of very small cookers and microwave ovens which plug directly into a 13 A socket outlet — these are regarded as 'portable appliances' for purposes of the IEE Wiring Regulations. Cooking appliances with a current rating exceeding 3 kW are regarded as 'fixed appliances' and should be supplied by their own final circuit (see Regulation 523–30). If, however, a situation arises where two cooking appliances, for example a split level cooker with built in oven and separate hob, are in the same room, then, provided each is within 2 m of the cooker control unit and the assumed current demand does not exceed 50 A, Regulation 476–20 states that they may both be controlled from the same switch (see Information Sheet No. 8A).

8.2 Control and protection of cooker circuits

The cooker final circuit must include a control switch or cooker control unit (CU) which can if desired contain a socket outlet (see Information Sheet No. 8A). The cooker final circuit is probably the largest of the final circuits used in the household situation and in Appendix 5 the IEE Wiring Regulations draw the attention of the installer to the need to afford discriminative operation of protective devices as stated in Regulation 533–6. As the local electricity board will normally use fuses to BS 88 for protection of their supply, this is usually interpreted to mean that this type of protective device should be employed for final circuits of this nature (see Chapter 3 for notes on discrimination).

Cooking and Space Heating 89

Information Sheet No. 8A Cooker circuits.

(a) Cooker unit

(b) Cooker circuit

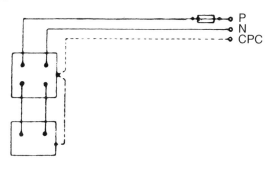

(c) Cooker circuit (separate hob)

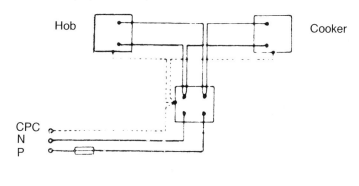

Thermostats

A thermostat is a device that will either open or close a set of contacts when a change in temperature takes place. They are used not only to control cookers, but heating and cooling appliances too, as well as being used to control space heating and ventilation systems.

The principle of operation is based on the rate at which different metals expand or contract due to changes in temperature. Strips of two different metals are fastened together (bi-metal strip) and when subjected to changes in temperature will bend due to their uneven expansion rates. The bi-metal strip is arranged in circuit to operate a set of contacts which will either be of the normally closed type (opening on temperature rise) or normally open (closing on temperature rise); the diagram on Information Sheet No. 8B shows an example of this.

Simmerstats

This useful device operates by opening and closing a switch at definite time intervals, the proportion of 'time on' and 'time off' being varied by turning a control knob. When connected to a hotplate circuit, it provides a gradual and varied means of control.

A bi-metal strip is surrounded by a heating coil and arranged to open and close a set of contacts. The control knob operates a cam to vary the setting. There is an arrangement to compensate for variations in temperature around the simmerstat itself (see Information Sheet No. 8B).

8.3 The application of diversity to cooker circuits

It is often reasonable to assume that not all the current-using apparatus connected to an electrical final circuit will be in use all of the time, therefore we can in certain circumstances make some allowance for this, resulting in the use of smaller cables, switchgear and protection which will cut the cost of the installation. Table 4A Appendix 4 of the IEE Wiring Regulations states that diversity may be applied to cooker final circuits used in the household situation and this is calculated as follows.

The first 10 A of the rated current plus 30% of the remainder of the rated current, plus 5 A if a socket outlet is incorporated in the control unit. An example of this is shown below.

If the total connected load of a cooker is 12 kW and the CU contains a socket outlet, what will be the anticipated maximum demand of the circuit after diversity has been applied?

Full load current:

$$P = I \times V$$

$$I = \frac{P}{V}$$

$$= \frac{12 \times 1000}{240}$$

$$= 50 \text{ A}$$

Cooking and Space Heating 91

Information Sheet No. 8B Simmerstats and thermostats.

1. Simmerstat

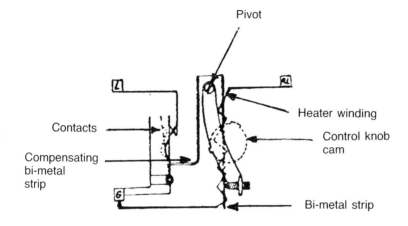

2. Thermostat

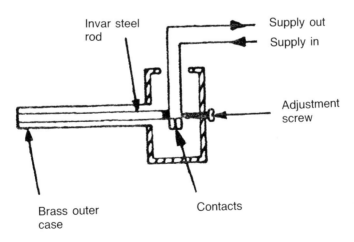

Anticipated maximum demand:

$$I = 10 + \frac{30 \times (50 - 10)}{100} + 5$$

$$I = 27 \text{ A}$$

It can be seen that the application of diversity can achieve quite a saving as a much smaller cable will be able to be used.

8.4 Heating

Heat transference

In all forms of heating, the object is to transfer heat energy from one body to another; from the heat source to its surroundings, or from one part of a substance to another. There are three different ways by which this transference can be effected, conduction, convection and radiation, as described below.

Conduction This consists of imparting heat by direct contact, for example through a substance from one part to another, or by two substances in contact with each other. A good example of this is the electric cooker ring which imparts its heat by conduction to the pan which in turn passes it to the food.

Convection This occurs when substances such as air, liquid, gas or vapour become warm due to a heat source. The heated substance expands and rises due to the decrease in its density, and is replaced by substance of a lower temperature. A good example of this is the convector heater, where cold air is drawn in through the lower grill, is heated by the electric element and rises to leave by the top grill (see Fig. 8.3). Cold air is once again drawn in to the bottom grill to replace the heated air and the process is repeated. The flow of air created in this way is known as 'convection currents' and these convection currents occur in all the substances mentioned.

Radiation This is a process that does not require a vehicle for transference of heat. It is the way the heat of the sun reaches earth. Heat is emitted from a heated body by radiation to other bodies of a lower temperature. However, the heat rays will pass through air or gaseous substances without heating them and will travel through a vacuum. 'Radiant Heat' which travels in straight lines is reflected by bright polished surfaces and absorbed by dark matt surfaces. Approximately 60% of the heat given off by an electric fire is due to radiant heat.

   Examples of electrical appliances using conduction, convection and radiation can be found on Information Sheet No. 8C.

Information Sheet No. 8C Conduction, convection and radiation.

1. Conduction

2. Convection

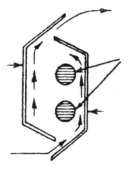

3. Radiation

8.5 Electric heating

Production of heat

When an electric current passes round a circuit the resistance of the conductor to the flow of current creates heat. This is called the 'heating effect' of an electric current. An electric fire has an element which is made of material having a high resistance, usually an alloy containing nickel and chrome (nichrome). When the fire is switched on this offers a great deal of resistance to the flow of current and the element heats up.

The classification of heaters

Electric heating can be classified under the following sub-headings:

- High temperature radiation, or radiant (including infra-red);
- Low temperature radiation (including block storage, floor-warming, ceiling and wall panels);
- Low temperature storage (including floor-warming cables);
- Low temperature convection (including tubular heaters, fan-heaters, etc.);
- Low temperature conduction (including heating tapes and soil-warming cables).

Types of heater

High temperature radiant heaters At one time the 'one bar electric fire' was probably the most common form of portable 'radiant heater' available. These may have several elements which consist of high resistance wire (usually an alloy containing nickel and chromium) mounted on a bar of refractory material. The bars can be selected by a switch so that the output can be raised or lowered at will. They are not normally thermostatically controlled; however, more elaborate ones suitable for use in low energy housing applications are available with this facility.

A number of radiant heaters are described as 'infra-red' space heaters. These consist of either the type based on a heating element, protected by a tubular metal or silica sheath mounted in front of an adjustable angle reflector, or the design based on the tungsten halogen lamp. Infra-red heaters are generally taken to be those which have been designed to emit radiant energy mainly at wavelengths within the infra-red portion of the spectrum. They make a safe, attractive and practical source of heat in situations where fast localised heating is required or for situations in which it would be impractical to maintain a comfortable air temperature. The element types are often incorporated into light fittings for use in bathrooms and kitchens, and the tungsten halogen types are often used for the heating of larger areas such as church halls, sports halls and warehouses.

Low temperature radiant heaters 'Low Temperature panels' consisting either of heating elements sandwiched between sheets of non-flammable material, or elements embedded in heat-resistant material, operate at temperatures ranging from 30°C to 65°C and can be incorporated within the ceiling structure. Panels of similar construction operating at higher temperatures may be used as wall mounted heaters. Installations of this kind, although basically low-temperature radiation, are not used for off-peak systems.

'Oil-filled radiators' resemble closely the sort of design usually associated with hot water central heating systems. They provide a versatile form of heating which is suitable for both domestic and commercial premises. Each radiator incorporates a thermostat which automatically responds to the temperature of the room, not the temperature of the radiator. This means the radiator will only use sufficient energy to maintain the temperature of the room at the chosen level. The radiators can be used from the normal switch socket outlet, though for bigger installations using a large number of heaters, specially designed circuits would be utilised. The oil is permanently sealed and under normal circumstances does not require changing or re-filling. The heating element is totally encased within the appliance, and most models incorporate a safety cut-out which ensures that the radiator does not overheat.

Low temperature storage heaters 'Thermal storage heaters' are quite simple in principle. A common unit of this kind consists of an electric heating element surrounded by refractory material, a layer of insulating material and a metal case. They are designed to be charged during the off-peak periods and then give out this heat slowly during the rest of the day. The slimmer models produced recently incorporate front and rear panels using microporous thermal insulation. The storage core is built up of very high density storage bricks, which have excellent thermal conductivity to release heat when required (see Fig. 8.1).

A charge regulator is often incorporated into thermal storage heaters which controls the amount of heat put into the core during the charge period. These can be adjusted from minimum to maximum charge and these

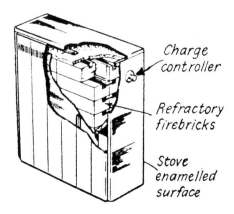

Fig. 8.1 Storage heater

may have the added facility of automatically adjusting for changes in weather. Boost controls may be used to release additional heat manually, or alternatively, set to do this automatically.

Combination heaters which combine a thermal storage heater and a thermostatically controlled convector heater in one unit are available. The integral direct acting convector heater provides the top-up heat when the storage heater component is not sufficient. By means of a room-temperature sensitive thermostat the convector heater can come into operation automatically to maintain a steady room temperature. The convector heater section has its own on/off switch and thermostatic control. To maximise the use of off-peak energy the convector heater incorporates a limit thermostat which is associated with an accelerator heater which comes on when the off-peak period commences (see Information Sheet No. 8D).

'Floor warming' simply consists of installing heating cables in the floor screed. To achieve a maximum temperature of 23°C to 29°C, loadings of 110 to 160 W per m^2 of floor area are usually sufficient. The cables are made of high resistance conducting material and insulated with mineral insulation, butyl or other heat-resisting insulating material. The cables are looped back and forward across the floor area and should be arranged to cover almost from wall to wall to achieve the best effect. The floor acts as a storage medium so this system of heating lends itself to off-peak heating tariffs such as 'economy seven'. It is a very efficient system and will only require supplementing with normal tariff heating in the coldest weather. Information Sheet No. 8D shows a diagram of a control system for thermal storage or floor-warming cable installations.

Low temperature convection heaters 'Tubular heaters' normally comprise round or oval metal tubes about 50 mm in diameter containing heating elements rated at 2 to 3 w/cm. They may be obtained in standard lengths from 30 cm to 500 cm, and may be arranged singly or in tiers. Applications range from frost protection to computer room heating (as no dust is created). They are ideal for domestic, horticultural or industrial premises and can be controlled by optional thermostats giving total economic control. The element of this type operates at black heat, giving a surface temperature of between 90 and 100°C. They are considered to be convector heaters, as very little heat is given up as radiation (see Fig. 8.2).

'Convector heaters' can be obtained as free-standing portable heaters or wall-mounting fixed appliances (see Fig. 8.3). For the most part they consist of a metal case with a grill at the top and at the bottom. Inside is an electric element, over which cold air, drawn through the bottom grill, passes. This air is heated and as explained earlier rises and leaves by the top grill so eventually heating the room. They can be controlled either by a 'three heat' switch, (see Information Sheet No. 8E) or they can be thermostatically controlled. Ideal for heating offices, workrooms, classrooms as well as the home, they come in various ratings from 0.5 kW to 3 kW.

'Fan heaters' can be regarded as a forced convection type of heater. An electric motor-driven fan behind a heating element gives an increased rate of air circulation, and has the added advantage that it can be used to blow cold air during hot weather. Portable types known as 'turbo-convector' are popu-

Information Sheet No. 8D 'Off peak' heating circuits.

1. Control system for thermal storage or floor warming cable installations

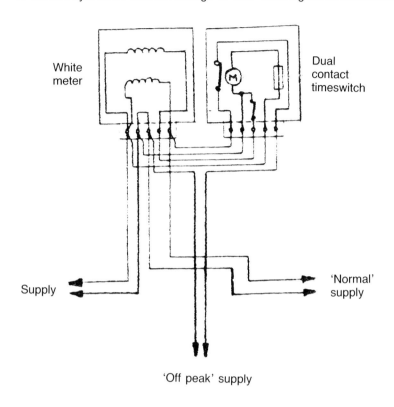

2. Limit thermostat

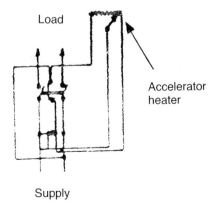

Information Sheet No. 8E Examples of 'three heat' switches.

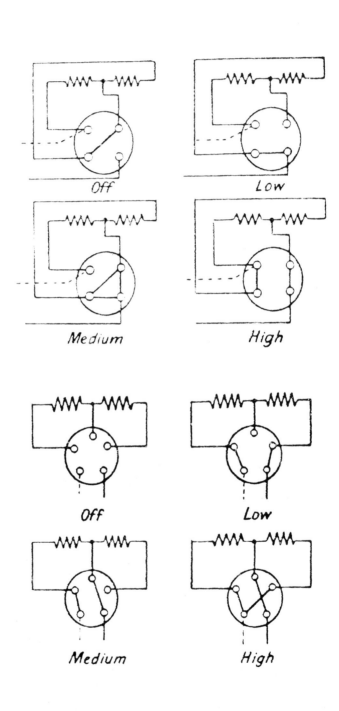

Cooking and Space Heating 99

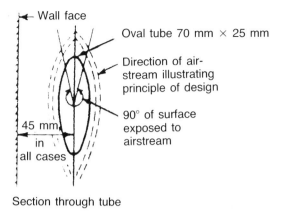

Fig. 8.2 Tubular heater

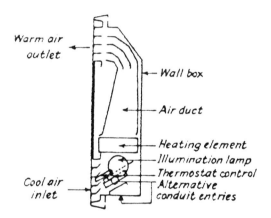

Fig. 8.3 Convector heater

lar, but for industrial and commercial use those described as unit heaters are generally used. Unit type fan heaters are generally provided with adjustable louvers to enable the warm air to be directed in the desired direction. They can be fixed to the wall or be mounted at high level, and they range in size from 3 kW to 20 kW, the larger ones often being of the 3 phase type.

Low temperature conduction heating 'Heating tapes' consisting of low-temperature high-resistance conductors embedded in heat-resistant plastic tape are used to good effect for the protection of pipes against frost. The tapes are installed close against the pipes under the lagging and the heat is conducted from the tape to the pipe-work and then ultimately to the water. These tapes can also be used with oil lines from oil storage tanks to the oil-fired boiler, keeping the viscosity of the oil at a suitable level.

'Soil-warming cables' are another example of how heat can be transferred by conduction. Much used by amateur gardeners, horticulturists and even football clubs, these cables are buried under the ground and the heat is conducted through the soil to plant or grass level providing the necessary heat for good growth and protection against frost. These cables consist of high-resistant low-heat conductors sheathed in heat-resistant plastic. The installer should be aware that in most cases when these are in use they will be outside the equipotential bonding zone and the IEE Wiring Regulations concerning this should be observed.

8.6 Space heating

Heating supplied to rooms in a building is usually known as space heating. To obtain maximum efficiency from electric space heating, thermal insulation should be installed throughout the building to reduce heat losses to a minimum.

For large buildings in the planning stage, advantages of direct electric heating are:

(1) no boiler rooms, chimneys, fuel stores or access roads thereto are necessary;
(2) moderate maintenance costs when compared to other heating systems;
(3) the installation can be more easily extended than other heating systems.

Space heating calculations

In space heating calculations, it is necessary to know the specific heat capacity of air (C) and also the mass of air in the room (m); mass is obtained by multiplying the volume of the room by the density of the air (1.28 kg/m^3). We also require to know the required number of air changes per hour (n) and temperature difference between temperature at commencement (t_1) and desired temperature (t_2).

Neglecting losses, the amount of heat energy (Q) in joules required to perform a space heating task is given by:

$$Q = m \times n \times C \times (t_2 - t_1)$$

Example The general office of a company measures 12 m × 9 m with a floor to ceiling height of 3.25 m. It has two doors of total area of 4 m^2 and a window down one side which has an area of 20 m^2. The heating is to be by electric convector heaters on normal tariff and are to maintain an inside temperature of 19°C when the temperature outside walls, ceiling and floor is 0°C. If there are to be two complete air changes per hour, calculate the required kW rating of the heaters.

Calculation of required energy (joules)

Volume of the room = 12 × 9 × 3.25 = 351 m^3
Changes per hour = 2
Density of air = 1.28 kg/m^3

Mass to be heated = 351 × 1.28 = 449.28
Specific heat capacity = 1010 (from Table 7.1)
Temperature rise = $t_2 - t_1$ = 19°C
$Q = m \times n \times C \times (t_2 - t_1)$
 = 449.28 × 2 × 1010 × 19
 = 17 243 366 J or 17 243.366 kJ

Calculation of electrical energy

We know from our previous work that the 'joule' is a common form of energy and that if a power of one watt is applied for one second then the electrical energy expended is one joule (watt second). This can be expressed as:

$$J = W \times s$$

therefore

$$W = \frac{J}{s}$$

So that for our example above the electrical energy expended is:

$$W = \frac{17\,243\,366}{3600}$$

 = 4789.82 W or 4.789 kW

N.B. This value ignores heat losses.

8.7 IEE Regulations concerning space heating

Many of the Regulations concerning the installation of final circuits previously discussed are applicable to the installation of circuits for electric heating. There are a number of Regulations, however, which apply to the installation of electric heating in particular and it might be worth looking at these.

Many of the heating appliances under 3 kW can be regarded as portable appliances, and may be plugged into the normal socket outlet in the domestic situation. However Appendix 5 of the Regulations makes it quite clear that permanently connected heating appliances forming part of a comprehensive space heating installation are to be supplied by their own separate circuit. Indiscriminate overloading can and does occur in commercial premises when portable heaters are plugged into socket outlets on standard circuit arrangements. This type of overloading occurs mainly when, for one reason or another, central heating systems of another type fail. People are tempted to plug in portable equipment in order to keep warm and overloading occurs.

A means of switching off for mechanical maintenance must be provided for equipment having electrically heated surfaces which can be touched (Regulation 476–7). The means of isolation shall be by a double pole linked switch which is either incorporated into the equipment (subject to Regulation 476–8), or is within easy reach of it. The switch should be marked to show when it is on or off and it might be found useful by the consumer if the switch contained a lamp to indicate this, especially if the equipment was of the non-luminous type.

Exposed flexible cable used for the final connection to fixed heating appliances must be kept as short as possible and termination to the fixed wiring of the final circuit shall be made with a suitable accessory or enclosure complying with Regulation 527–4 or, where Chapters 43 and 46 of the Regulations so require, a suitable device for overcurrent protection, isolation and switching.

The type and current-carrying capacity of cables and flexible cords used for the connection of appliances shall be suitable for operation in the highest temperature likely to occur in normal service (Regulation 523–1).

In rooms containing a fixed bath or shower no stationary appliance having heating elements that can be touched shall be installed within reach of a person having a bath or shower and this includes elements sheathed in silica glass (Regulation 471–39).

Switches and controls operating the appliance shall be out of reach of people using the bath or shower unless operated by an insulated pull-cord or unless they are part of SELV circuits (Safety Extra Lower Voltage) complying with Regulation 471–39 (a).

No electrical equipment of any sort must be installed in the interior of a bath or shower basin and there shall be no socket outlets or provision for connection of portable appliances (Regulation 471–34).

For circuits supplying equipment in rooms containing a fixed bath or shower all the requirements for bonding etc. referred to in Regulation 471–35 should be met. If the equipment is simultaneously accessible with exposed conductive parts of other equipment or extraneous conductive parts (for definitions of these see Part Two of the Regulations), then the protective device should be capable of disconnecting the circuit in 0.4 s if an earth fault occurs.

Fixed heating equipment shall be selected and erected so that its intended heat dissipation is not inhibited and does not present a fire hazard to adjacent building materials. Any equipment which has a normal operating temperature on its surface of over 90°C must be adequately ventilated, and any building material that presents a fire hazard must not be placed nearer than 300 mm above this equipment or 150 mm laterally or below it (see Regulation 422–1 and 2). These distances can be reduced if a suitable fire-resistant material is used to shield or enclose any such material. Appliances installed to the manufacturers recommendations and complying with BS 3456, which concerns itself with the safety of domestic and similar appliances, are considered to be exempt from this regulation.

Where under-floor or soil-warming cables come into close proximity of materials which present a fire hazard, Regulation 554–31 recommends that they should be enclosed in material that complies with BS 476 Part 5 with an ignitability characteristic 'P' and shall be adequately protected against mechanical damage.

The loading of floor-warming cables shall operate at temperatures not exceeding those listed in Table 55D of the IEE Regulations.

Test 8

Choose which of the four answers is the correct one.

(1) The IEE Wiring Regulations regard any cooking appliances with a current rating exceeding 3 kW as a:

(a) portable appliance;
(b) radial circuit;
(c) fixed appliance;
(d) heating load.

(2) A split level oven and separate hob unit can both be connected to the same cooker control unit provided that:

(a) each are within 6 m of the control unit;
(b) they have their own isolation switch;
(c) they are connected in series;
(d) each are within 2 m of the control unit.

(3) Heating supplied to rooms in a building is usually known as:

(a) specific heat;
(b) space heating;
(c) selective heating;
(d) forced convection.

(4) Circuits supplying equipment placed in a room containing a fixed bath or shower shall be capable of:

(a) automatic disconnection within 0.4 s;
(b) supplying 3 kW of power;
(c) automatic disconnection within 5.0 s;
(d) supplying 6 kW of power.

(5) Switches in rooms containing a fixed bath or shower shall not be in reach of a person using the bath or shower, unless:

(a) operated by an insulated pull cord;
(b) made entirely of plastic;
(c) operated by a rocker switch;
(d) engraved with the word 'heater'.

Chapter 9
Electric Lighting

9.1 The incandescent lamp

Lighting was one of the first major applications for the use of electricity, Sir Joseph Swan producing the first lamp in 1880. This consisted of a carbon strip in a vacuum tube, and produced a very poor light. By 1910 metal filament lamps had come into being, they were more robust and the temperatures that they operated at could be taken higher so producing more light. More recent developments have seen the use of gas-filled lamps which allow operating temperatures in excess of 2700°C. Most of the lamps used today have filaments made from tungsten, manufactured in the form of a spiral (see Fig. 9.1).

The lamp filament offers resistance to the flow of electric current, generating heat and causing it to incandesce (glow with heat) so producing light. The greater part of energy consumed is given off in the form of heat and the light output is biased towards the infra-red end of the spectrum. The lamp emits radiation covering the visible spectrum, and is much used for general lighting in domestic and commercial premises being of low cost and maintenance free. This type of lamp has an efficacy (lumens per watt) range 10 to 18 lm/W.

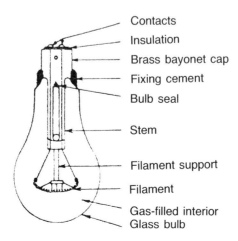

Fig. 9.1 Incandescent lamp

9.2 The tungsten filament lamp

Types of tungsten filament lamp

Twelve distinct sizes make up the 'General Lighting Service' (GLS) range of lamps, 15, 25, 40, 60, 75, 100, 150, 200, 300, 500, 750 and 1000 W, all these being made for a maximum voltage of 250 V. 'Rough service lamps' are available for use on building construction sites, industrial machine lighting etc. Their reinforced internal construction gives increased resistance to filament breakage due to jolts and vibrations. Variations on the basic lamp are as follows.

Vacuum lamps Tungsten filament in a single coil, enclosed in a glass envelope (bulb) with a complete vacuum inside. Has temperature limitations and is confined to use for general purpose low wattage lamps. Evaporation of the tungsten filament over a period of time tends to leave a black deposit on the inside of the lamp.

Gas-filled Tungsten filament spiralled into a single coil, enclosed in a glass envelope filled with gas. The gas used is argon, nitrogen or a mixture of the two, this cools the filament and allows it to run at a much higher temperature. The light given off is whiter than the vacuum lamp and it has a greater efficacy.

Coiled coil In this lamp the filament is coiled a second time, thus producing a more compact filament design. This reduces heat loss and allows greater efficacy for the same temperatures.

Obscure To reduce glare, and to some extent shadow, which are a feature of the above lamps, manufacturers produce a lamp whose envelope has a pearl finish. This alleviates the problems and gives very nearly the same light output. Further diffusion can be obtained by the use of lamps with an internal coating of silica, both types being ideally suited for domestic use.

Internal reflector Where it is desired to direct the light output of the lamp or concentrate the beam, the interior of the lamp is mirrored to act as a reflector.

Functional envelope design Glass envelopes can be obtained in many shapes and sizes to suit the particular function that the lamp will be required to serve. These vary from the pressed glass robust construction of the 'par reflector lamps' much used in display lighting, to the delicate flickering candle lamp used in chandeliers.

Lamp cap design The majority of tungsten filament lamps have caps of either the Bayonet Cap (BC) type or the Edison Screw (ES) type. These lamps can be used in any position but they are designed to operate most efficiently cap uppermost. There are a number of other lamp cap designs as shown on Information Sheet No. 9A.

Information Sheet No. 9A Common British Lamp cap designs.

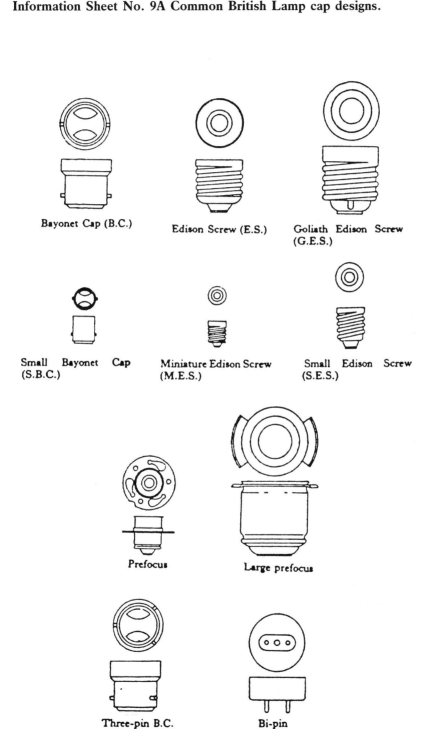

Variation in voltage

Running a lamp at a lower voltage than that of its rating, results in the light output of the lamp being reduced at a greater rate than the electricity consumption and the lamp's efficacy is poor. This reduction in voltage, however, increases the lamp's life and can be useful where lamps are difficult to replace or light output is not the major consideration.

Running a lamp at a higher voltage than it has been designed to do results in a shorter lamp life. It has been calculated that by an increase of just 5% the lamp's life can be halved. This is done deliberately with 'Photoflood' lamps used in photographic work, where the light output is more important than lamp life, just 1% over rating producing 3.5% increase in lamp output (lumens). When one considers, however, that Electricity Supply Authorities are allowed to vary their voltage up to and including 6% it is easy to see that if this was carried on for any length of time the lamps would not last very long.

9.3 The tungsten halogen lamp

These lamps were introduced in the 1950s. For their operation the tungsten filament is enclosed in a gas-filled quartz tube together with a carefully controlled amount of halogen such as iodine or bromine. When the filament is heated by an electric current the tungsten is evaporated from the surface of the filament and is carried by convection currents to the comparatively cool walls of the lamp. Here it combines with the halogen which has vaporised and forms a tungsten halide. This compound returns to the filament where the high temperatures chemically convert it back to tungsten, the halogen gas that is left sinks, to be drawn again through the filament in a continuing cycle.

Because the tungsten is returned to the filament the blackening of the walls of the lamp does not take place like it does with the GLS lamp. A minimum bulb wall temperature of 250°C is needed to maintain the cycle and therefore a smaller bulb can be utilised. This allows much higher gas pressure so that the life of the halogen lamps can be up to 2000 hours, double that of the tungsten filament lamp. The material used for the construction of the bulb or tube is quartz, which in comparison with glass has very little thermal expansion and it can operate at higher temperatures resulting in a whiter light.

There are four halogens; iodine, fluorine, chlorine and bromine. Iodine is used commercially for lamps of long life, fluorine which is corrosive, has been used for experimental lamps and bromine is used in lamps for the lighting of television studios. Lamps available at the moment are described as the quartz–iodine, tungsten–iodine or the tungsten–halogen lamps, although to date most commercially produced lamps use iodine as the halogen.

The advantages of the tungsten halogen lamp are:

- Increased lamp life;
- 100% lumen output maintained;

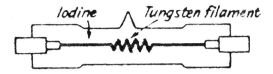

Fig. 9.2 Tungsten halogen lamp (double ended)

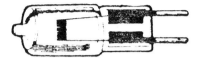

Fig. 9.3 Tungsten halogen lamp (single-ended)

- Increase in luminous efficacy (21 lm/W);
- Reduction in lamp size;
- Reduction in luminaire size and therefore cost;
- Improvement in optical performance of fittings used.

To date two basic designs have been produced:

(a) a double-ended linear lamp with the contacts in the seal at each end (see Fig. 9.2);
(b) a single-ended lamp with both contacts embedded in the seal at one end (see Fig. 9.3);

The linear lamp must be operated within 4° of the horizontal to prevent the halogen vapour from migrating to one end of the tube causing early failure. The lamps must be handled very carefully especially when being fitted. If the outside of the quartz tube is contaminated with grease from the hand, fine cracks will appear in the surface of the tube causing premature failure. In practice it is advisable to leave the paper wrapping round the lamp until it is fitted in place or alternatively handle it only by the ends. If accidental contact is made then it should be cleaned with a solvent such as industrial spirit or carbon tetrachloride.

9.4 Fluorescent lighting

Principles of operation

The fluorescent lamp, or low pressure mercury vapour lamp to give it its correct name, consists of a glass tube filled with a rare gas such as krypton or argon with a measured amount of mercury vapour. Coated on the inside of the glass tube is a phosphor or rare earth and in each end there is sealed a set of oxide-coated electrodes. Because these electrodes are heated it is sometimes referred to as the hot cathode lamp (see Fig.9.4).

Unlike the fittings described previously which rely on incandescence for their light output, the fluorescent lamp is a form of discharge lamp. That is

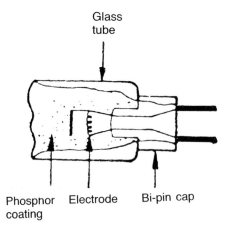

Fig. 9.4 Fluorescent lamp

to say it relies on a discharge taking place across the length of the lamp between the two electrodes. Before this discharge can take place in the lamp, its gas filling must be ionized. The voltage to carry out this ionization must be much higher than the voltage required to maintain the actual discharge across the lamp and the manufacturers use several methods to achieve this high voltage usually based on a transformer or choke.

The discharge is at the ultra-violet end of the spectrum so would be of little value to us as a light source; however, it causes the phosphors coated on the inside of the tube to fluoresce and give off visible light, hence the name fluorescent tube. By employing different phosphors, different colour renderings can be obtained.

A look at any of the fluorescent circuits will show that once the discharge has taken place, the lamp is virtually connected directly across the mains supply. This could be dangerous because the impedance across the lamp gets less and less as the discharge establishes itself and the current gets higher and higher. It is at this point that the secondary function of the choke takes place; as the current increases the impedance of the coil in the choke increases, limiting the current across the lamp and keeping it in balance, one reason why it is often referred to as a ballast.

Unfortunately the high inductance of the coil in the choke or ballast creates a bad 'power factor' (pf), often as poor as 0.5 lagging, so that a pf correction capacitor must be included in the circuit to compensate for this. There are a number of different ways that these circuits can be arranged and some of the more common ones are shown on Information Sheet No. 9B.

The glow starter circuit In the starter, normally open contacts are mounted on bimetal strips and enclosed in an atmosphere of Helium gas. When switched on, a glow discharge takes place around the open contacts in the starter, which heats up the bimetal strips causing them to bend and touch each other. This puts the electrodes at either end of the fluorescent tube in circuit and they start to warm up giving off electrons; at the same time an intense magnetic field is building up in the choke which is also in circuit.

Information Sheet No. 9B Fluorescent circuits.

Glow-starter switch circuit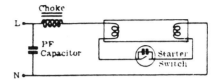

Thermal starter-switch circuit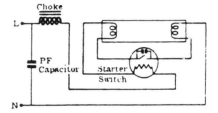

Quickstart Circuit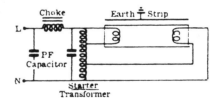

Semi-resonant start circuit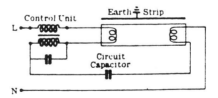

Twinstart circuit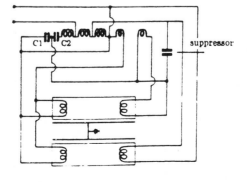

The glow discharge in the starter ceases once the contacts are touching, so that the bimetal strips now cool down and they spring apart again. This breaks the circuit causing the magnetic field in the choke to collapse and providing the momentary high voltage required for ionization of the gas and enabling the main discharge across the lamp to take place. The voltage across the tube under running conditions is not sufficient to operate the starter and so the contacts remain open. The resistance of the mercury vapour gets less as it gets warmer (negative temperature coefficient) and conducts more current. This could lead to disintegration of the tube; however, the choke has a stabilising effect as described earlier. This type of starting method may not succeed first time and gives recourse to the characteristic flashing on switching on.

Quickstart circuits (transformer type) The electrodes in this circuit are rapidly pre-heated by the end windings of an auto-transformer so that a quick start is possible. When the electrodes reach the required temperature the capacitive effect between the lamp cathodes and the earthed metal work of the luminaire ionises the gas in the tube; some lamps have an earthing strip along the length of the tube to assist in this. Since the transformer has a high reactance compared to the choke in circuit, most of the mains voltage appears across the transformer and therefore across the lamp and is sufficient to start the discharge once the electrodes are ready. When the lamp is alight the voltage across it and the transformer drop to a lower value, consequently the heating current supplied by the transformer end windings also drops. This circuit will 'strike' the lamp first time; however, it does have the disadvantage that if the the mains voltage is low it may be reluctant to start, and it also relies heavily on the earthing arrangement.

Thermal starter circuit It must be said that this type of circuit has lost its popularity in recent years. However, as there are thousands of these fittings still in service, it will be worth describing it.

In this starter, normally closed contacts are mounted on a bimetal strip. One of these is heated by a small heater coil when the supply is switched on. This causes the strip to bend and the contacts open causing the momentary high voltage described earlier and so starting the discharge. The heater is in circuit even when the lamp has 'struck' so that if it becomes open circuit the lamp will not operate. The starter is easily recognised as it has four pins on it instead of the usual two, the extra pins being for the heater connections.

Semi-resonant circuit In this fitting the place of the choke is taken by a specially wound transformer. On switching on, current flows through the primary winding of the transformer, through one of the electrodes of the tube, then back through the secondary winding of the transformer which is wound in opposition to the first one. A capacitor is in circuit between the secondary winding and the second electrode of the lamp, the other side of which is connected to neutral so completing the circuit.

The current which flows through the circuit rapidly heats the electrodes and as the circuit is predominantly capacitive, the pre-start current leads the mains voltage. Owing to the fact that the primary and secondary windings are in opposition to each other, the voltages developed across them are 180° out of phase with each other, so that the voltage across the tube is high, causing the tube to 'strike'. The current builds up across the lamp; however, the primary winding which is still in circuit after the lamp strikes, acts as a choke stabilising the circuit. This circuit has the advantage of a high power factor due to the circuits capacitor and has the ability to start in cold weather because it is not temperature dependent as other circuits. As in the case of the quick start fitting the earthing arrangement is important.

Electronic ballast circuits An electronic ballast takes advantage of the fact that fluorescent lamps operate more efficiently at greater frequencies. The electronic ballast operates in the 20 to 60 kHz range and the use of an electronic pre-regulator ensures low harmonic distortion at input, the ability to withstand normal mains surges and gives a power factor near unity. The number of circuit components is reduced to a single all electronic solid state unit which is silent in operation, has instant starting and eliminates flicker.

Circuit Diagrams and associated control gear are shown on Information Sheet No. 9B.

9.5 Stroboscopic effects

A simple illustration of this effect is seen when you visit the cinema, particularly if there is a western showing. How many times while watching this type of film have you noticed the wheels of the wagon appearing to be stopped or even going backwards. This phenomenon is brought about by the fact that the spokes of the wagon wheel are rotating at or about the same revolutions per second as the frames per second of the film that is being shot. We can get the same effect from a 'stroboscope', which is a lamp that has a flashing light, if we shine it on a revolving piece of machinery. By adjusting it until the periodic flashing matches the revolutions of the machine we can get it to appear as if the machine was stationary, just like the wagon wheels.

Unfortunately fluorescent luminaires can under certain circumstances have the same stroboscopic effect on moving machinery with potentially dangerous implications. For example a four-spoked wheel revolving at 1500 rpm will have one of its spokes at the 12 o'clock position 100 times every second. At the same time the fluorescent lamp, due to the fact that it is on AC supply at a frequency of 50 Hz, will have its discharge extinguished 100 times every second (twice per cycle). This will result in the spoke appearing to be stationary.

This effect must be taken into consideration when contemplating the installation of this type of luminaire. There are a number of things that we can do to overcome the problem:

(1) We can install a tungsten filament lamp fitting locally to the machine; many lathes can now be seen with lamps mounted as standard on the machine. The reason for this is that the tungsten filament is so hot that it does not have time to cool down during the AC cycle and so does not have this effect on rotating machinery.

(2) Adjacent fluorescent fittings can be connected to different phases of the supply. Because in a three phase supply the phases are 120° out of phase with each other the light falling on the machine will arrive from two different sources. Each of these will be flashing at a different time which will interfere with each other and spoil the stroboscopic effect.
(3) Twin fittings can be installed with the lamps wired on lead lag circuits thus counteracting each other.
(4) One of the recently developed 'high frequency' fluorescent fittings described earlier can be installed. The frequency is so high that this type of fitting does not produce the effect.

Some of the fluorescent lamps, particularly the warm colours, and other types of discharge lamps which we will deal with in Book 3 suffer from the stroboscopic effect to a lesser extent. This is because the phosphors used are phosphorescent as well as fluorescent and the 'after glow' continues for a short while after the discharge is extinguished. Thus the lamp when used on the normal 50 Hz AC supply is still giving off some light even at the instant of zero lamp current.

Maximum demand current

The current used by a discharge lighting circuit cannot be obtained simply by dividing the lamp watts by the voltage. This is because of losses in the control gear, low power factor and harmonics present in the supply current. For this reason if no information is available from the manufacturer then the maximum demand current can be calculated by taking the lamp watts and multiplying by a factor of 1.8. This is provided that the power factor is no worse than 0.85. Unless the switches used on these circuits have been specially designed for use on discharge lighting, they shall have a nominal current of not less than twice the steady current which they are required to carry.

9.6 High voltage discharge lighting

Principles of operation

Often referred to as 'neon' lighting, although several different gases are in fact used in this sort of lighting.

The lamp consists of a glass tube often bent to take the form of a letter or pattern. This is filled with low-pressure neon gas or a mixture of gases to obtain the different colours. Unlike the fluorescent tube described earlier it is a cold cathode lamp, relying on the high voltage applied across the tube to ionise the gas and sustain the discharge. A double wound step up transformer is used to obtain the high voltage and this is centre tapped to earth. The highest permissible voltage to earth is 5000 V AC so by the use of a centre tapped transformer a 10 000 V AC output is obtained (see Information sheet No. 9C).

Electric Lighting

Information Sheet No. 9C High voltage discharge lighting.

1. High voltage discharge lighting circuit

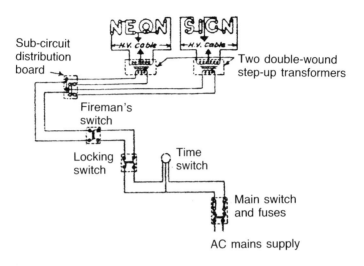

2. Warning labels

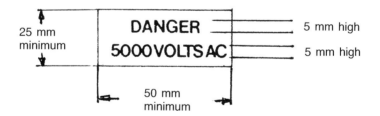

Regulations concerning high voltage discharge lighting

Apart from the smallest self-contained signs the installation of high voltage discharge lighting is a highly specialised job and BS 559 Part 3 lays down the requirements for its installation as detailed below.

Transformers – control and protection

Auto transformers can be used in place of the double wound transformer mentioned above, but must not exceed 1500 V r.m.s. measured on open circuit, must be fed from a supply with one pole earthed (TN or TT) and be controlled by a double poled isolator.

Transformers with an input rating greater than 500 W shall be provided with a protective device which shall provide automatic disconnection of both poles of the supply in the event of short circuit or accidental contact between any part of the high voltage circuit and earth. The rated operating current shall not exceed 10 mA and means shall be provided to verify the operation. A residual current operated device is not considered suitable, since when connected on the primary side of the transformer it does not protect against faults to earth on the secondary side.

Enclosures

The transformer and ancillary equipment shall be totally enclosed in an earthed metal enclosure, which can form part of the sign or luminaire. Otherwise it must be in a ventilated enclosure constructed from material having the ignitability characteristic 'P' as specified in BS 476 part 5. Any openings or apertures in the enclosure must comply with classification IPX 4 of BS 5490 which must be capable of excluding objects 1 mm thick or 1 mm in diameter (terminal screwdriver).

Warning signs

Enclosures containing transformers or ancillary equipment which are accessible to unauthorised persons shall have safety signs placed upon them. These will conform to A.2.8 of BS 5378 Part 1 1980, which takes the form of a triangle with a lightening zig zag symbol upon it. Under this shall be another sign saying 'DANGER HIGH VOLTAGE' in black coloured letters 5 mm high on a yellow background, the whole thing not less than 50 mm by 25 mm overall. Alternative wording can be the word danger with the highest voltage stated (see Information Sheet 9C). Any removable covers shall be fastened in such a way as to require a tool to remove them.

IEE Regulations concerning high voltage discharge lighting

The requirements for the installation of high voltage discharge lighting is now out of the scope of the IEE Wiring Regulations, these having come under BS 559 since 1986. However the wiring of the low voltage (above 50 V but not exceeding 1000 V AC) final circuits, isolation and the provision of 'Fireman's Switches' clearly does come within the scope.

Fireman's switch

The requirements for the fireman's switch are clear: it must be capable of isolating all live conductors and be clearly marked ON and OFF and this should be legible to a person standing on the ground. The 'off' position shall be at the top and preferably with a lock or catch to prevent it being inadvertently returned to the 'on' position (see Information Sheet No. 9D). It should be coloured bright red and should have on it or near it a nameplate saying FIREMAN'S SWITCH . The name plate should be the minimum size of 150 mm by 100 mm, in lettering legible from a distance depending on the site but in any case not less than 13 mm high. It is desirable that the name of the company who installed or who are responsible for the maintenance of the installation be included on the notice (see Regulation 537–17).

The positioning of the fireman's switch is important. Its purpose is to provide emergency isolation of high voltage discharge lighting installations for both exterior installations and interior installations which run unattended. Therefore the switch should be placed in a conspicuous position, reasonably accessible to firemen and mounted not more than 2.7 m from the ground. For exterior installations the switch will be outside the building and adjacent to the high voltage discharge lamp installation. Alternatively a permanent notice indicating the location of the switch shall be fixed at the installation and another notice fixed where the switch is located clearly indicating its purpose.

For interior installations the switch shall be installed in the main entrance to the building. In both cases, where difficulty is experienced with the siting of the switch the positions can be varied after consultation with the local fire officer or the responsible person in the local authority. If there is more than one switch for an installation each switch should clearly indicate the area that it controls, but wherever possible installations on any one building should be controlled by a single switch; however, separate switches should be used for interior and exterior installations (see Regulation 476–12 and 13).

Isolation of high voltage discharge lighting

Further to the requirements for isolation and protection of high voltage discharge lamp installations mentioned in BS 559 Part 3 above, the IEE Wiring Regulation 476–6 states that the isolator concerned must be in addition to the switch normally used to control the circuit. It is recommended that devices used for isolation are selected and/or installed so as to prevent inadvertent and unauthorised operation. This can be achieved by locating the isolator in a lockable space or enclosure, by padlocking, by removal of the isolator's handle or by the use of a lockable distribution board. It is

Information Sheet No. 9D Switchgear for high voltage discharge circuits

1. Fireman's switch

2. Warning notice
   not less than
   150 mm × 100 mm

3. Lockable switchfuse

4. Lockable isolator

Electric Lighting   119

important to ensure that keys or handles for the above are not interchangeable with those of other parts of the installation. Examples of lockable switchfuses and isolators suitable for this purpose are shown on Information Sheet No. 9D. Self-contained luminaires can have an interlocking device fitted so that they must be disconnected from the supply before access to live parts can be obtained (see Regulation 476−4).

9.7 Emergency lighting

Secondary lighting systems that automatically come on when the mains electricity has failed are highly desirable for the safety of the occupants of many different types of premises. While this is useful under any circumstances, their prime function is to enable people to evacuate the building safely in an emergency. Some regulations now require their installation in such premises as hotels, boarding houses and places of public entertainment. The Fire Precautions Act lays down the general provisions, but this can vary a little from one part of the country to the other, often depending on the interpretation of the local Fire Officer. Even when there is no legal requirement the public can still be at risk, and the installation of such a system should be a serious consideration when the electrical installation design of a building is being considered.

Regulations governing emergency lighting

Because the majority of emergency lighting schemes are fed from a safety source as detailed in IEE Wiring Regulation 411−3 they do not come within the scope of the IEE Wiring Regulations. However, any low voltage supplies associated with the installation do, and it is desirable that the standards set for this should be maintained for the rest of the installation. In general the fundamental requirements are laid down in BS 5266: Part 1. This provides guidance on the location of emergency luminaires in hazardous places and the illumination of the escape route. In March 1986 the Chartered Institute of Building Services Engineers (CIBSE) produced TM 12, a publication on emergency lighting which gives excellent guidance on lighting level requirements, and the Industry Committee for Emergency Lighting (ICEL), an association of some of the leading manufacturers of emergency lighting products, set standards for the manufacture of luminaires and equipment.

The escape route is defined in BS 5266: Part 1 as 'a route forming part of the means of escape from a point in the building to the final exits'. The final exit is the point 'beyond which persons are no longer in danger'. Once the hazardous points and the escape route has been established, thought will have to be given to the lighting levels that need to be maintained. The recommended levels are 0.2 lux on the centre line of a clearly defined escape route which is to be a maximum of 2 m (or strips of 2 m for wider routes), with 50% of the centre area required to be illuminated to at least 0.1 lux. There are more details for fixed seating, movable seating and stepped areas. A short duration minimum of 2 lux is recommended for locations where

120 Electrical Installation Practice 2

special tasks might have to be carried out for safety reasons before evacuation. Examples of this might be turning off gas supplies or shutting down plant and or machinery. Planning data based on various heights for luminaires and different wattage fittings are available, whilst most of the manufacturers produce helpful charts and ready reckoners giving examples of layouts for luminaires. It should be stressed that these should only be used as a guide and are no substitute for a qualified engineer.

Type of system

Selection of the most suitable system will generally be a question of economics. There are two main options open:

(a) Central battery system powering slave luminaires located throughout the building (see Fig. 9.5).
(b) self-contained emergency luminaires each with its own power unit consisting usually of nickel-cadmium cells (see Fig. 9.6).

In the smaller building, self-contained units are likely to be more economic. They are also easier to maintain, requiring virtually no attention other than a duration test every six months or as stipulated by the local Fire Officer. As a general guide installations involving, say, more than 25 luminaires are likely to be more economically served by a central battery system, though this would have to be determined after looking at the layout drawings for the particular job and determining the lengths of cable runs.

Self-contained luminaires have certain technical advantages. It is generally easier to extend this type of system, as each unit is simply connected to the public supply wiring at the nearest point. They will remain on charge until that particular circuit fails, so if there is a fire, each individual fitting will

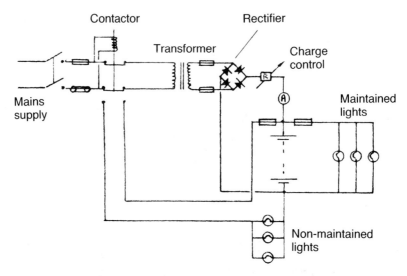

Fig. 9.5 Emergency lighting central battery system, A=ammeter, R=variable resistor

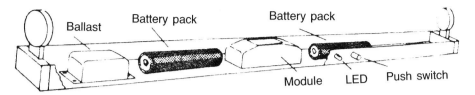

Fig. 9.6 Self-contained emergency luminaire

continue to function until the fire reaches that point. The luminaires fed by a central battery system would all fail once the power source was enveloped. Advantages of the central battery system are that the luminaires are relatively inexpensive compared to the self-contained ones and the choice of fitting design is almost unlimited. Disadvantages of the self-contained fittings include the cost of replacement cells, which is very high particularly on a large installation, and as the average length of life of the nickel-cadmium cell is five years, this becomes an important consideration. In addition to this the choice of luminaire is limited to the ones which the manufacturers have chosen to utilise; although most of the manufacturers will consider using a luminaire of your choice if requested, it can be expensive. Figure 9.6 shows a fluorescent luminaire fitted with nickel-cadmium cells.

Maintained, non-maintained or sustained

These are sometimes confusing terms describing three operations in emergency lighting systems.

Maintained — a system in which the emergency luminaire remains illuminated all the time, even when the mains supply is functioning. The same lamp is utilised whether it is being supplied via a transformer/rectifier arrangement from the mains supply, or from an extra low voltage safety source.

Non-maintained — systems which do not have their luminaires illuminated until such time as the normal lighting fails. It is essential therefore that they are tested in accordance with BS 5266 to ensure there are no failures.

Sustained — systems having two lamps in each luminaire. One lamp is powered by the mains supply and the other by the emergency supply. A typical use for this type of fitting is the exit signs used in places of public entertainment.

The choice of these three will depend on local regulations. Non-maintained are the lowest in cost. A circuit diagram showing a typical arrangement for a central battery system serving both 'maintained' and non-maintained' operations is shown in Fig. 9.5.

Battery duration

Batteries for both the central system and self-contained fittings can be specified with varying durations. The most common are one hour and three hours – the former being long enough to evacuate most buildings in an emergency, the latter for larger complexes. The normal duration assumes that the battery is fully charged when the emergency circuit is activated. Frequently the local authorities determine the duration required for emergency lighting, leaving the designer to specify suitable equipment.

Test 9

Choose which of the four answers is the correct one.

(1) Most of the 'incandescent lamps' used in recent times have filaments made from:

(a) carbon;
(b) nickel-chrome;
(c) tungsten;
(d) nickel-cadmium.

(2) The initials BC written on a packet containing a lamp, indicate that it is a:

(a) Bare Cap types;
(b) Bi-pin Cap;
(c) Bi-pin Connection;
(d) Bayonet Cap type.

(3) In the 'tungsten halogen' lamps the most common halogen to be used in their manufacture is:

(a) iodine;
(b) fluorine;
(c) neon;
(d) quartz.

(4) The secondary function of the transformer or choke in a fluorescent circuit is to:

(a) improve the power factor;
(b) limit the current in the circuit;
(c) cause the electrodes to glow;
(d) limit the voltage across the lamp.

(5) Another name for the fluorescent lamp is the:

(a) high pressure mercury lamp;
(b) low pressure mercury lamp;
(c) cold cathode lamp;
(d) H.V. discharge lamp.

Chapter 10
The Inspection and Testing of Installations

10.1 Inspecting and testing

Requirements of the IEE Wiring Regulations

Every electrical installation shall be inspected and tested in accordance with the Regulations before being connected to the public supply. This is to ensure as far as practicable that all the requirements of the Regulations have been carried out and the installation is safe to use. The Regulations require that the tests carried out shall not in any way be a danger to persons, property or equipment, even if the circuit is faulty. It is important then that the tests are carried out in the recommended sequence shown in Part 6 of the Regulations and this is as follows:

1. Ring final circuit continuity.
2. Protective conductor continuity (including all bonding).
3. Measurement of earth electrode resistance.
4. Insulation resistance.
5. Insulation of site built assemblies.
6. Protection by electrical separation.
7. Protection by barriers and enclosures.
8. Insulation of non-conducting floors and walls.
9. Verification of polarity.
10. Earth fault loop impedance.
11. Operation of residual current devices.

10.2 The testing of installations

The testing of the Continuity of The Ring Final Circuit Conductors, Continuity of Protective Conductors and Equipotential Bonding, Insulation Test, Polarity Test and the Visual Inspection of the installation were dealt with in Book 1 of the series. This section deals with the remaining tests and also looks at insulation testing again but this time from the point of view of large installations.

Earth electrode resistance

Make sure that the electrode is disconnected before commencing the test. The test is carried out by passing a steady value of AC through the electrode

T and an auxiliary electrode T1 which is placed at such a distance from T that the resistance areas of the two do not overlap. A second auxiliary electrode T2 is then inserted halfway between T and T1, the voltage drop between T and T2 is measured and noted. The resistance of the earth electrode is then the voltage between T and T2, divided by the current flowing between T and T1. To check that this is a true value the electrode T2 is moved 6 m from and 6 m nearer to T respectively. If the readings are more or less in agreement then the mean of the three readings can be taken as the resistance of the earth electrode (see Fig. 10.1).

Insulation resistance

The purpose of the insulation test is to ensure that the insulation is sound and that no faults exist between phase and neutral conductors and between each of these conductors and earth. The test is carried out with the circuit to be tested isolated from the mains supply, using an insulation tester on which the megohm scale has been selected. The voltage used must be twice that of the supply, but need not exceed 500 V DC for installations rated up to 500 V, or 1000 V DC for installations above 500 V.

When testing between phase and neutral make sure that all lamps have been removed, and that all appliances are either unplugged or isolated from the circuit by switching them off. The fuses must be in place and all switches in the on position (other than ones isolating appliances from the circuit). When these conditions have been satisfied a reading is taken and this must not be less than one megohm (one million ohms).

When testing between phase and earth and neutral to earth, it is common practice to twist phase and neutral together and test between these and earth. The instrument used is the same as for above and the reading taken in megohms. Isolate the supply as near to the mains intake position as possible,

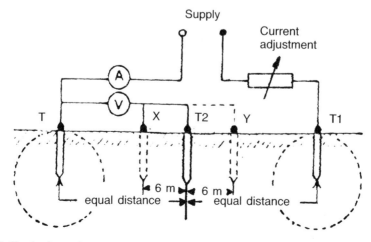

Fig. 10.1 Earth electrode test

make sure the fuses are in place and any breakers and switches are in the on position. When these conditions have been met a reading is taken and this must not be less than one megohm. If a fault should be detected, it will be necessary to test between phase and earth and neutral and earth separately, in order to ascertain which of these conductors the fault is on (see Information Sheet No. 10A).

For three-phase and neutral systems the requirements for fuses, switches, isolators and fixed equipment are the same as for single-phase supplies. However, this time each of the four conductors will require to be tested to see that the insulation resistance between each of them and the other conductors is not less than one megohm and that the insulation resistance between each of them and earth is also not less than one megohm.

When carrying out insulation tests on large installations, IEE Wiring Regulation 613-5 states that the installation can be split up into groups of not less than 50 outlets for the purposes of the tests outlined above. This includes every point and every switch, except that socket outlets and luminaires containing switches are counted as one. The reason for this is that when circuits are connected together in parallel the results can be misleading. For example three circuits with readings of two megohms when tested individually, would result in a reading of 0.666 megohms when tested together.

Where equipment has been disconnected in order to carry out the tests, if it is practical the equipment itself must undergo an insulation test. The test result must comply with the BS Standard for the equipment; if, however, there is no standard the insulation resistance shall be not less than 0.5 megohm.

Insulation of site built assemblies

Where protection against direct contact with live parts of a site built piece of equipment is by insulation applied during erection, Regulation 412-2 says that it should completely cover the part and be able to withstand any mechanical, electrical, thermal and chemical stresses to which it may be subjected in service. The quality of the insulation should be confirmed by tests similar to those which ensure the quality of insulation of factory built equipment, details of which can be found in BS 5486.

Electrical separation of circuits

Safety can be achieved in special situations by the use of electrical separation. The mains supply is taken to the primary side of an isolating transformer, while the unearthed secondary feeds the equipment to be protected. Common examples of this are the shaver sockets (to BS 3052) for rooms containing a bath or shower, or the isolation transformers often used in laboratories.

Further details can be found in Regulations 413-35 to 39.

The Inspection and Testing of Installations

Information Sheet No. 10A Insulation testing.

1. Testing between phase and neutral and the CPC.

2. Testing between the phase and neutral conductors.

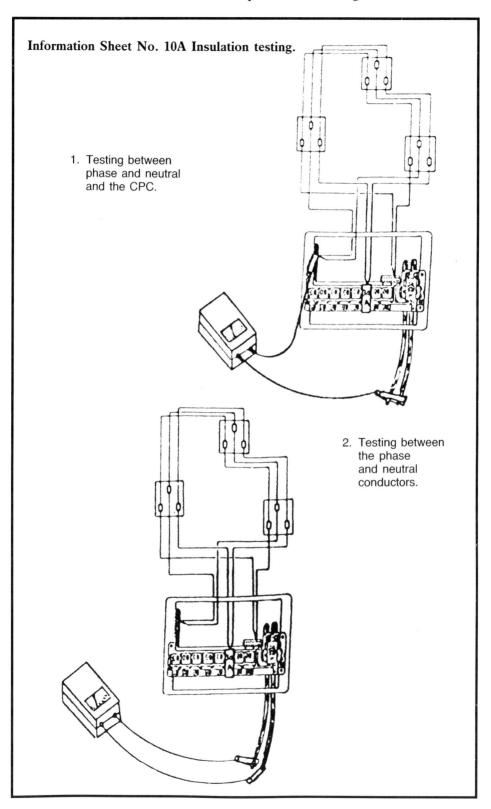

Protection by barriers or enclosures

Regulations 412–3 to 412–6 and Regulation 613–13 refer to protection by barriers or enclosures to BS 5490, to prevent contact with live parts. The protection for horizontal top surfaces of barriers will be IP 4X classification, designed so as to prevent a probe of more than 12 mm diameter and 80 mm long (the average human finger) from making contact with the live parts. Other protection should be to IP 2X classification, designed to prevent probes greater than 1 mm diameter or 1 mm thick (terminal screwdriver) from coming into contact with live parts.

Insulation of non-conducting floors and walls

Protection can be afforded by placing exposed conductive parts in a non-conducting location. The exposed conductive parts shall be arranged so that under ordinary circumstances a person will not come into simultaneous contact with either two exposed conductive parts, or an exposed conductive part and an extraneous conductive part, if these parts are liable to be at different potentials due to failure of the basic insulation. See Regulation 413–27 to 413–31 (a).

The resistance between the walls and floor of such a location and the main protective conductor of the installation shall be measured at not less than three points on each of the surfaces in question. One of these points shall be not less than 1 m and not more than 1.2 m from any extraneous conductive part in the location. See Regulation 613–13 for further details.

Earth fault loop impedance test

These tests are carried out using an instrument which measures the current flowing when a known resistance is connected between the consumer's earth terminal and the phase conductor, as shown in Information Sheet No. 10B. Before commencing with this test it is important that the continuity of the protective conductor is confirmed. A close look at the diagram will show that if it was incomplete in any way this would result in the whole of the protective system being connected directly across the phase conductor. This is one of the reasons why tests should be carried out as close to the recommended sequence as possible.

It is important that the impedance of the earth fault loop path is low enough to enable the supply voltage to pass a current high enough to operate the protective device of the circuit and in the time stipulated in the IEE Wiring Regulations. This will of course vary in accordance with the type of protective device used and its current rating. Minimum impedance values can be found for different types of protection in Tables 41A1 and 41A2 of the Regulations.

Information Sheet No. 10B Residual current devices (RCD) and earth impedance tests.

1. RCD test

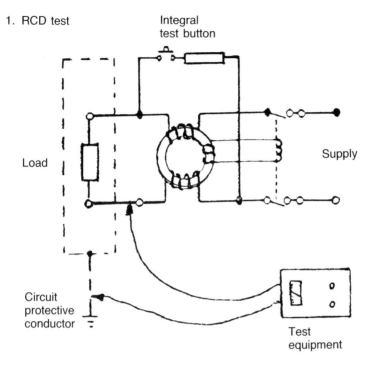

2. Earth impedance test

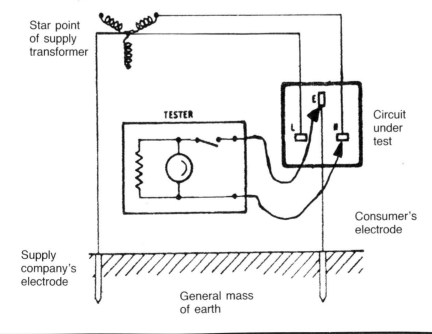

Operation of residual current devices

Where the earth fault loop impedance is of such a value that overcurrent protective devices cannot operate within the stipulated disconnection times mentioned in Regulation 413−4, then with the exception of circuits incorporating PEN conductors, residual current devices (RCD) are used. Although RCDs have built in test facilities, the Regulations require that a separate test be carried out to ensure their satisfactory operation under fault conditions.

The test instrument usually consists of a single phase, double wound, step-down transformer with an output voltage of approximately 45 V and voltmeter. The test is made on the load side of the circuit breaker between the phase conductor of the circuit protected and the associated circuit protective conductor so that a suitable residual current flows. All the loads normally supplied by the circuit breaker are disconnected during the test.

The rated tripping current shall cause the circuit breaker to open within 0.2 s or at any delay time declared by the manufacturer of the device. Where the circuit breaker has a rated tripping current not exceeding 30 mA and has been installed to reduce the risk associated with direct contact a residual current of 150 mA should cause it to operate within 40 ms. The latest test sets simply plug into a socket outlet to carry out the test, but in no event must the test current be applied for longer than one second.

10.3 Certification

Inspection certificate

An inspection certificate should be given by the Electrical Contractor or person acting on his behalf to the client on completion of the inspection and testing of an installation. A typical example of this certificate is shown in Appendix 16 of the IEE Wiring Regulations. The certificate certifies that the inspection and testing of the electrical installation has been carried out in accordance with the IEE Wiring Regulations (the current edition) and that the results are satisfactory. Any poor results or non-compliance with the Regulations must be noted in the spaces provided. This is particularly important when testing an existing installation, as there might be a number of departures from the current edition of the Regulations, and it is the responsibility of the testing engineer to bring these to the client's notice.

Completion certificate

When the installation has been inspected and tested in accordance with the relevant tests listed above, and the faults and omissions if any rectified, then a completion certificate should be issued to the client or his/her representative. It should take the form of the example shown in Appendix 16 of the IEE Wiring Regulations; if your company belongs to any of the organisations such as the National Inspection Council for Electrical Installation Contracting (NICEIC), then these organisations will provide you with copies for your use. It will be seen that the certificate should contain full details of the completed installation, such as the number of lighting points, socket outlets, fixed equipment etc. Any departures from the IEE Wiring Regulations in the design and execution of the installation should be noted and these should have had approval of a qualified electrical engineer. The certificate verifies in effect that the installation is completed and ready for use and should be signed by a qualified electrical engineer, a member or qualified representative of the Electrical Contractors Associations of England and Wales or Scotland, or Approved Contractor of the NICEIC. The certificate should be handed over together with any relevant drawings etc. and a copy of the Inspection Certificate.

Periodic testing and inspection

IEE regulation 514–5 asks that a notice be fixed in a prominent position on or near to the Main Distribution Board of every completed electrical installation, indicating that the installation should be periodically inspected and tested. The notice should give the date of the last inspection and details of when it should be tested again. The period between tests should be five years for normal installations, but this can be made less if this is thought appropriate. Agricultural installations should be tested every three years, and temporary installations on construction sites every three months. Caravan or mobile homes should be inspected and tested every year, but this can be extended up to three years where appropriate.

Test 10

Choose which of the four answers is the correct one.

(1) Before conducting an insulation test the following should be carried out:

(a) all lamps removed and appliances unplugged or disconnected;
(b) all lamps and equipment to be used are placed in circuit;
(c) the insulation tester is switched to the ohms scales;
(d) the insulation tester is of the 1000 V AC type.

(2) When carrying out an insulation test on a large installation it can be split into groups of:

(a) not less than 20 outlets;
(b) not less than 50 outlets;
(c) lighting and power circuits;
(d) high and low voltage circuits.

(3) 'Electrical separation' can be achieved by:

(a) dividing the circuits into categories 1, 2 or 3;
(b) keeping the neutral separate from the cpc conductor;
(c) the use of double insulation;
(d) the use of an isolating transformer.

(4) Before carrying out an earth fault loop impedance test, it is important to confirm that the:

(a) polarity of the test leads is correct;
(b) the supply is 240 V AC;
(c) continuity of the neutral conductor has been established;
(d) continuity of the protective conductor has been established.

(5) When testing the operation of residual current devices the test is made on the load side of the RCD between the:

(a) phase conductor and associated neutral conductor;
(b) neutral conductor and associated circuit protective conductor;
(c) phase conductor and associated circuit protective conductor;
(d) earth electrode and associated neutral conductor.

Answers to the Tests

Test 1

(1) (c); (2) (a); (3) (d); (4) (b); (5) (c).

Test 2

(1) (c); (2) (b); (3) (c); (4) (a); (5) (d).

Test 3

(1) (d); (2) (c); (3) (b); (4) (d); (5) (a).

Test 4

(1) (b); (2) (c); (3) (d); (4) (a); (5) (c).

Test 5

(1) (b); (2) (c); (3) (d); (4) (d); (5) (a).

Test 6

(1) (c); (2) (c); (3) (a); (4) (b); (5) (d).

Test 7

(1) (c); (2) (a); (3) (d); (4) (b); (5) (c).

Test 8

(1) (c); (2) (d); (3) (b); (4) (a); (5) (a).

Test 9

(1) (c);   (2) (d);   (3) (a);   (4) (b);   (5) (b).

Test 10

(1) (a);   (2) (b);   (3) (d);   (4) (d);   (5) (c).

Index

Ambient temperature, 19
Animals (damage), 20
Architect, 1
Assignments, 7
Automatic disconnection, 36

Bayonet cap (BC), 107
Bench and dado trunking, 66
Block diagrams, 8
BS 3939 symbols, 11
Bus-bar chamber, 41
Butyl, 19

Cable tray, 52
Cable trunking, 61
Cartridge fuse, 45
Categories (cables), 74
Central battery systems, 120
Certification, 130
Choke, 110
Circuit diagrams, 8
Circuit protective conductors, 32
Class 2 equipment, 27
Clerk of Works, 2
Client, 1
Company structure, 4
Conduction, 92
Conductors, 17
Consultant, 1
Contractual relationships, 4
Control and protection, 40
Convection, 92
Cooker final circuits, 88
Correction factors, 24
Corrosive substances, 20

Design current, 22
Design team, 2
Designers, 1
Detached buildings, 43
Direct contact, 27
Discrimination, 47
Distribution, 40
Distribution board, 41
Diversity (cookers), 90

Earth, 27
Earth clip, 33

Earth electrodes, 33
Earth fault loop impedance, 36
Earth potential, 31
Edison screw, 107
Electric cables, 17
Electric heating, 84
Electrical separation, 27
Electrolytic action, 20
Electromagnetic effect, 75
Emergency lighting, 119
Equipotential bonding, 29
Equipotential bonding zone, 31
Exposed conductive part, 31
Extraneous conductive parts, 31

Fireman's switch, 117
Floor joists, 22
Fluorescent lighting, 109
Flush floor trunking, 68
Foreman, 7
Fusing factor, 46

General mass of earth, 27

HBC fuse, 45
Heating, 92
HV discharge lighting, 114

Incandescent lamp, 105
Indirect contact, 27
Inspection and testing, 124
Installation team, 6
Insulation, 17

Ladder racking, 59
Layout drawings, 8
Lighting, 105
Lighting trunking, 65

Magnesium oxide, 19
Main contractor, 3
Maximum demand current, 114
Measuring, 10
Mechanical damage, 20
Miniature circuit breaker, 46
Moisture, 20

Nominated sub-contractor, 3

Nominated supplier, 3
Non-conducting location, 29

Off peak supplies, 96
Overcurrent, 44
Overhead bus-bar trunking, 71

Phenol-formaldehyde, 19
Planning, 8
Plastic trunking, 73
Polyvinyl chloride, 17
Professional relationships, 4
Protection against shock, 27
Protective device, 23

Quantity surveyor, 2

Radiation, 92
Requisition sheet, 14
Residual current device, 37
Resistivity, 18
Rising main trunking, 73

Scale (drawings), 10
Segregation, 74
Selection (cables), 22
Self contained luminaires, 120
Shaver socket BS 3052, 27
Simmerstats, 90
Site supervision, 7

Skirting trunking, 66
Space factor, 62
Space heating, 100
Specialist, 3
Star point, 28
Stress, 20
Stroboscopic effect, 113
Sub-contractor, 3
Sunlight (direct), 20
Supplementary bonding, 31
Suppliers, 3
Suspended floor installations, 70
Suspension (cables), 21
Switchgear, 43

Take off sheet, 13
Thermosetting polymer, 19
Thermostats, 90
Three heat switch, 98
TN-C-S system, 37
TN-S system, 37
TT system, 37
Tungsten filament lamp, 106
Tungsten halogen lamp, 108

Voltage drop, 23

Water heating, 77
Wiring diagrams, 8